Walid Kaakeh

Global Reporting Initiative (GRI) in Promoting Sustainability

Walid Kaakeh

Global Reporting Initiative (GRI) in Promoting Sustainability

A Comprehensive Guide to Frameworks, Practices, and Global Impact for Transparent and Sustainable Business Reporting

Noor Publishing

Cover image: www.ingimage.com

Publisher:
Noor Publishing
is a trademark of
Dodo Books Indian Ocean Ltd. and OmniScriptum S.R.L publishing group

120 High Road, East Finchley, London, N2 9ED, United Kingdom
Str. Armeneasca 28/1, office 1, Chisinau MD-2012, Republic of Moldova, Europe
Managing Directors: Ieva Konstantinova, Victoria Ursu
info@omniscriptum.com

Printed at: see last page
ISBN: 978-620-8-87066-9

Global Reporting Initiative (GRI) in Promoting Sustainability

A Comprehensive Guide to Frameworks, Practices, and Global Impact for Transparent and Sustainable Business Reporting

by

Prof. Dr. Walid Kaakeh
International Consultant

First Edition – 2025

بسم الله الرحمن الرحيم

IN THE NAME OF ALLAH
THE MOST BLESSED, THE MOST MERCIFUL

Dedication

To my mother and father,
Aisha and Abdul Ghani Kaakeh
May God bless them and make them dwell
in His spacious gardens

To my wife
Fatima Khatib
for her unwavering patience, compassion, and most
importantly her love

To my beloved children,
Tasneem, Yasmine, Yamen, and Enas
who continue to be a source of inspiration and pride

I dedicate this book

Walid Kaakeh

Table of Contents

Comprehensive Introduction

In an era defined by unprecedented environmental, social, and economic challenges, the concept of sustainability has emerged as a cornerstone of global development. Organizations worldwide are increasingly recognizing the need to balance profit with purpose, ensuring that their operations contribute positively to society and the planet. At the heart of this transformation lies sustainability reporting—a practice that enables organizations to transparently communicate their environmental, social, and governance (ESG) performance. Among the various frameworks and standards available, the Global Reporting Initiative (GRI) stands out as a pioneering force in promoting sustainability reporting. This book, Global Reporting Initiative (GRI) in Promoting Sustainability, delves into the multifaceted role of GRI in shaping sustainable business practices and fostering accountability across industries.

The journey of sustainability reporting has evolved significantly over the past few decades. From its early days as a voluntary practice to its current status as a critical component of corporate governance, sustainability reporting has become a vital tool for organizations to demonstrate their commitment to sustainable development. Chapter 1, Introduction to Sustainability Reporting, provides a comprehensive overview of this evolution, highlighting the definition, importance, and historical context of sustainability reporting. It also explores the emergence of reporting standards and their significance in driving transparency and comparability across organizations.

At the core of this book is the Global Reporting Initiative (GRI), a global standard-setter that has revolutionized the way organizations report on their sustainability performance. Chapter 2, Founding and Objectives of the Global Reporting Initiative (GRI), traces the history and establishment of GRI, shedding light on its mission and vision. Since its inception in 1997, GRI has been instrumental in developing a universal framework for sustainability reporting, enabling organizations to measure, manage, and communicate their impacts on critical sustainability issues. This chapter underscores GRI's role as a catalyst for change, empowering organizations to align their strategies with the principles of sustainable development.

One of the key strengths of GRI lies in its robust reporting frameworks and guidelines. Chapter 3, GRI Reporting Frameworks and Guidelines, provides an in-depth analysis of the GRI Standards, which serve as the foundation for sustainability reporting. It also explores the Core and Comprehensive Option frameworks, offering insights into their applicability and relevance for organizations of varying sizes and sectors. By adhering to these frameworks, organizations can ensure that their sustainability reports are comprehensive, credible, and aligned with global best practices.

The benefits of GRI reporting extend far beyond compliance. Chapter 4, Benefits of GRI Reporting, examines the organizational and stakeholder advantages of adopting GRI standards. For organizations, GRI reporting enhances reputation, improves risk management, and drives innovation. For stakeholders, including investors, customers, and communities, GRI reporting provides valuable insights into an organization's sustainability performance, fostering trust and accountability. This chapter highlights the transformative potential of GRI reporting in creating shared value for businesses and society.

Implementing GRI reporting, however, is not without its challenges. Chapter 5, GRI Reporting Process, outlines the steps involved in adopting GRI standards, from initial planning to final publication. It also emphasizes the importance of materiality assessment, a critical process that enables organizations to identify and prioritize the sustainability issues that matter most to their stakeholders. By integrating materiality into their reporting processes, organizations can ensure that their sustainability reports are relevant, focused, and impactful.

The alignment of GRI with the United Nations Sustainable Development Goals (SDGs) is another key theme explored in this book. Chapter 6, GRI and Sustainable Development Goals (SDGs), examines how GRI reporting supports the achievement of the SDGs. By providing a structured framework for measuring and reporting on sustainability performance, GRI enables organizations to contribute to the global agenda for sustainable development. This chapter also highlights the synergies between GRI and SDG reporting, illustrating how organizations can leverage GRI standards to advance their sustainability goals.

Stakeholder engagement is a cornerstone of effective sustainability reporting. Chapter 7, GRI and Stakeholder Engagement, underscores the importance of involving stakeholders in the reporting process. It also presents best practices for engaging stakeholders, ensuring that their perspectives and concerns are reflected in sustainability reports. By fostering meaningful dialogue with stakeholders, organizations can enhance the credibility and relevance of their reports, building stronger relationships with their audiences.

Assurance plays a critical role in enhancing the credibility of sustainability reports. Chapter 8, GRI and Assurance, explores the types of assurance available for GRI reporting and the benefits and challenges associated with each. By undergoing external assurance, organizations can demonstrate their commitment to transparency and accountability, reinforcing stakeholder trust in their sustainability performance.

The concept of integrated reporting, which combines financial and non-financial information, is gaining traction as a means of providing a holistic view of organizational performance. Chapter 9, GRI and Integrated Reporting, examines the synergies between GRI and integrated reporting, highlighting how organizations can leverage GRI standards to enhance their integrated reports. This chapter also explores the potential of integrated reporting to drive sustainable value creation, offering insights into its growing relevance in the corporate world. As sustainability reporting continues to evolve, GRI has developed industry-specific standards to address the unique challenges and opportunities faced by different sectors. Chapter 10, GRI and Industry-Specific Reporting, provides an overview of sector supplements and specific standards, illustrating how organizations can tailor their reporting to reflect the realities of their industries. This chapter also includes examples of industry-specific reporting, showcasing the diverse applications of GRI standards across sectors.

Small and medium enterprises (SMEs) play a vital role in the global economy, yet they often face unique challenges in adopting sustainability reporting practices. Chapter 11, GRI and Small and Medium Enterprises (SMEs), explores how GRI standards can be adapted for SMEs, highlighting the benefits and challenges of sustainability reporting for smaller organizations. By embracing GRI reporting, SMEs can enhance their competitiveness, build stakeholder trust, and contribute to sustainable development.

Climate change is one of the most pressing challenges of our time, and organizations have a critical role to play in addressing it. Chapter 12, GRI and Climate Change Reporting, examines GRI's guidance on climate change reporting, including the measurement and reporting of carbon footprints. This chapter underscores the importance of transparent and accurate climate change reporting in driving organizational action and supporting global efforts to combat climate change.

Human rights and gender equality are fundamental principles of sustainable development, and GRI standards provide a framework for incorporating these issues into sustainability reporting. Chapters 13 and 14, GRI and Human Rights Reporting and GRI and Gender Equality Reporting, explore how organizations can integrate human rights and gender equality indicators into their reports. These chapters also include case studies and examples, illustrating how organizations are using GRI reporting to promote social justice and equality.

Technology innovation is transforming the way organizations approach sustainability reporting. Chapter 15, GRI and Technology Innovation, examines the role of technology in enhancing the efficiency and accuracy of GRI reporting. It also explores emerging trends in technology and reporting, offering insights into how organizations can leverage digital tools to improve their sustainability performance.

The circular economy is gaining momentum as a model for sustainable resource management, and GRI standards provide a framework for reporting on circular economy performance. Chapter 16, GRI and Circular Economy Reporting, explores the concepts and principles of circular economy reporting, highlighting the performance indicators that organizations can use to measure their progress. This chapter also includes examples of circular reporting, showcasing how organizations are embracing the circular economy to drive sustainability.

Supply chain transparency is a critical issue for organizations seeking to enhance their sustainability performance. Chapter 17, GRI and Supply Chain Reporting, examines the role of GRI standards in promoting supply chain transparency and supplier engagement. This chapter underscores the importance of responsible supply chain management in achieving sustainability goals, offering practical guidance for organizations seeking to improve their supply chain reporting.

Biodiversity and water stewardship are essential components of sustainable development, and GRI standards provide a framework for reporting on these issues. Chapters 18 and 19, GRI and Biodiversity Reporting and GRI and Water Reporting, explore how organizations can integrate biodiversity and water stewardship into their sustainability reports. These chapters also include metrics and case studies, illustrating how organizations are using GRI reporting to address these critical issues.

Anti-corruption is a key aspect of corporate governance, and GRI standards provide a framework for reporting on anti-corruption measures. Chapter 20, GRI and Anti-Corruption Reporting, examines how organizations can incorporate anti-corruption measures into their sustainability reports. This chapter also includes case studies, showcasing how organizations are using GRI reporting to promote ethical business practices.

The concept of circular reporting, which emphasizes the continuous improvement of sustainability performance, is explored in Chapter 21, GRI and Circular Reporting. This chapter examines the principles and examples of circular reporting, highlighting its potential to drive organizational change and enhance sustainability performance.

The debate between voluntary and mandatory reporting is a key issue in the sustainability reporting landscape. Chapter 22, GRI and Voluntary vs. Mandatory Reporting, explores the benefits and challenges of voluntary reporting, as well as the trends towards mandatory reporting. This chapter also examines the implications of these trends for organizations and stakeholders, offering insights into the future of sustainability reporting.

The adoption of GRI standards in emerging markets is a critical issue for global sustainability. Chapter 23, GRI and Reporting in Emerging Markets, examines the challenges and opportunities associated with the adoption of GRI standards in developing economies. This chapter also highlights the potential of GRI reporting to drive sustainable development in emerging markets, offering practical guidance for organizations seeking to enhance their sustainability performance.

The Non-Financial Reporting Directive (NFRD) is a key regulatory framework for sustainability reporting in the European Union. Chapter 24, GRI and Non-Financial Reporting Directive (NFRD), explores the harmonization of GRI standards with the NFRD, highlighting the implications for organizations and stakeholders. This chapter also examines the potential of the NFRD to drive global convergence in sustainability reporting, offering insights into the future of reporting regulation.

Finally, Chapter 25, GRI and Future Trends in Reporting, explores the innovations and predictions shaping the future of GRI reporting. This chapter examines the emerging trends and technologies that are transforming the reporting landscape, offering insights into how organizations can stay ahead of the curve in their sustainability reporting practices.

In conclusion, Global Reporting Initiative (GRI) in Promoting Sustainability provides a comprehensive exploration of the role of GRI in driving sustainable development. By offering practical guidance, case studies, and insights into emerging trends, this book serves as an invaluable resource for organizations seeking to enhance their sustainability performance and contribute to a more sustainable future. Through the adoption of GRI standards, organizations can demonstrate their commitment to transparency, accountability, and sustainable development, building a better world for future generations.

Chapter 1
Introduction to Sustainability Reporting

Review: This section introduces sustainability reporting, its definition, and its importance in modern business practices.

Key Topics: Definition of sustainability reporting, its role in transparency, accountability, and decision-making.

Outcome: Chapter 4 demonstrates that GRI reporting creates significant value for organizations and stakeholders. For organizations, it enhances decision-making, operational efficiency, and competitive advantage by providing standardized sustainability metrics. For stakeholders, it fosters transparency, accountability, and informed decision-making, building trust and engagement. The chapter highlights GRI's role in reducing information asymmetry, promoting global sustainability standards, and driving both internal and external value creation.

1.1. Introduction

The use of the guidelines as a tool in corporate sustainability reporting is apparently widespread. A recent survey of corporate sustainability reports from corporations in various Emerging Asian economies found that most reports cite the use of the guidelines (Romana, 2018). However, the application of the sustainability guidelines may be less widespread, or at least less mentioned, in smaller economies, as is the case with the sustainability reporting in the Philippines, which is only cited by one of the four corporations examined.

The Guidelines also take into account the reporting of environmental, economic and social performance. These include, among others, the "to go where the needs are", the "fair operating practices" (safety practises in the work, environmental protection, and social responsibility), and employment creation, including vocational training and benefit for local communities, these being common features of extractive companies. The focus of the Corporate Social Responsibility (CRS) activities can be on “community involvement and development”. A range of problems has been alleged by environmental pressure groups against the TNC, including the significant gas leaks in the early phases of the offered exploration activities, that the operations pose a significant threat to the delicate regions, the lack of public knowledge on the activities due to insufficient disclosure, and that corruption and bribery are generating instability in the region.

1.2. Definition and Importance of Sustainability Reporting

Review: Explores the concept of sustainability reporting and why it is critical for organizations.

Key Topics: Transparency, stakeholder trust, and long-term value creation.

Outcome: Highlights how sustainability reporting drives organizational accountability and stakeholder confidence.

In recent years, the sustainability of the business field has gained a massive interest all over the world. Since the descriptive approach of sustainability, businesses are expected to reveal the effect of their operations on social and environmental issues. This issue was the source of the birth of the global sustainability initiative and the inclination to publish sustainability reports. The Global Reporting Initiative provides sustainability standards for corporations and aims to attract as many corporations as possible to the sustainability reporting framework (Hysa et al., 2015). By revealing transparency on economic, natural, and social affects of the reporting companies, the declaration tries to set the cornerstone for understanding and communicating the importance of observing sustainable development standards. - Reporting practices are maturing enterprises that discard the burden of "voluntary" reporting and are adapted to considerations of available resources, presentation capabilities, expertise, or even better governance. For fictional reporting of the companies about the financially related issues that have been interacting throughout the centuries, the trend in accordance with the GRI guidelines for sustainability matters through a chronicled era remained generally latent.

Sustainability Reporting as an Activity in the Corporate World has Lately Encountered Attention in Various Enterprise Areas (Wuerzinger, 2015). Within Academic Circles, the Focus on the Theme Has Seeable Progress But no Evaluation Framed over Epoka Universities, Which is the Case of Discussion in the Current Paper or Developed Countries. It is also Assumed that for qualifying to discuss environmental issues and to take a chance to discuss social implications, a prescribed or evaluated assessment report must be established. Is Known the Plan of Office of University X for the formulation of GRI-related objectives, sanctioned under the label of the Project for Realization of Center of Sustainability. The main goal is to deepen the rights of expanding the University's default setting instrument in sustainability matters on the national plane. A best way is to evaluate opportunities about higher education institutions is Epoka Universities as a case in the discussion. A Center on Sustained Mission will be equipped to the University's stakeholder, which will be charged with the task of drafting the University's annual sustainability report. The objective of the paper is thus to discusses the instruments, objectives, task, and prospects of ECOSSCence to be established and on the consequence of the relevant investments in regard to a case in point.

1.3. Evolution and Significance of Reporting Standards

Review: Traces the historical development of reporting standards and their impact.

Key Topics: Emergence of global standards, standardization benefits, and comparability.

Outcome: Demonstrates the importance of standardized frameworks in enhancing reporting quality and consistency.

Today's sentiment is rising excitement due to expectations for more disclosed information, higher comparability across companies and increased frequency of publication of sustainability reports. A widely ratified framework for sustainability reporting with its comprehensive guidelines that underwent several transformations during time and have now been used extensively by companies worldwide. Reporting standards usually consist of reporting principles, content elements and the method of measurement – they help reporters in preparation and assurance in discouraging missing, vague, misrepresented or irrelevant information in the reports. The sustainability reporting guidelines are non-prescriptive, voluntary guidelines to be used when preparing a sustainability report. The aim of these standards is to increase the amount and quality of disclosed sustainability information that will affect the performance of guidelines has been increasing significantly since they were established in 1997. Initiating of own sustainability projects, improved sustainability performance, enhanced use of sustainability reporting as a management tool, satisfaction of stakeholder expectations were observed as effects of the widespread adoption of the guidelines beyond the introduction year 1997, nearly a thousand companies and NGOs were using the CER since 2000.

Launched G2 Guidelines in 2000, there was a substantial increase in the number of sustainability reports being produced worldwide, however, criticism of the G2 Guidelines in terms of being difficult to use, being applied too rigidly, the credibility of companies' reports and the credibility of the itself led to gradual decline in use of the guidelines until the G3 Guidelines were introduced. With the transitioning from the G2 to the G3 guidelines, there is a notable difference in the quantity of reports received by a new . Some countries implemented regulations for sustainability reporting. Despite the efforts of this multi-stakeholder initiative, however, ever-growing demands for higher quality of reported sustainability information remained unfulfilled with the G2 Guidelines and there was a noticeable decline in the number of sustainability reports issued by the companies using the G2 Guidelines in years before those guidelines were renounced. Prompted by the criticism of G2 Guidelines and hoping to open a new era of sustainable development, the revised the outdated G2 Guidelines resulting in a wide consultation process and introduced the G3 Guidelines in 2006 (Wuerzinger, 2015). Prior to 2006, however, the annual publication of sustainability reports using the guidelines went into a rapid decline, the impact of the guidelines fairly quickly faded out as well as there are hardly any sustainability reports produced by firms. Although numerous other factors contributed to this very large global decline in the quantity and quality of sustainability reporting and thereby significantly increased the environmental and socio-political risks of firms, the final G2 more than others is detrimental to the fate of the CER in Iberia. G3 Guidelines proved to be far more successful and thus promote the sustainability in Spain and Portugal. G4 Guidelines are not a one-size-fits-all standard framework but rather a more flexible set of standards that can be adapted to a company's needs. There are a few reasons why the G4 Guidelines is a marked improvement over the preceding versions of the Guidelines.

1.4 Challenging Questions and Answers

Q1: Why is sustainability reporting considered critical for modern businesses, and how does it contribute to stakeholder trust?

A1: Sustainability reporting is critical because it provides transparency into a company's environmental, social, and economic impacts. By disclosing this information, businesses demonstrate accountability and commitment to sustainable practices, which builds trust among stakeholders such as investors, customers, employees, and communities. Trust is fostered through clear communication of efforts to address sustainability challenges, such as reducing carbon emissions, improving labor practices, or contributing to community development.

Q2: What are the key challenges faced by smaller economies in adopting sustainability reporting frameworks, as highlighted in the chapter?

A2: Smaller economies often face challenges such as limited resources, lack of expertise, and insufficient awareness of sustainability reporting frameworks. For example, the chapter notes that in the Philippines, only one out of four corporations examined cited the use of sustainability guidelines. This suggests that smaller economies may prioritize immediate economic concerns over long-term sustainability goals, and there may be a lack of regulatory pressure or incentives to adopt these practices.

Q3: How have the GRI guidelines evolved over time, and what improvements were introduced in the G4 Guidelines compared to earlier versions?

A3: The GRI guidelines have evolved significantly since their inception in 1997. The G2 Guidelines faced criticism for being rigid and difficult to use, leading to a decline in their adoption. The G3 Guidelines, introduced in 2006, addressed some of these issues but still had limitations. The G4 Guidelines, however, marked a significant improvement by offering a more flexible and adaptable framework. They allowed companies to tailor their reporting to specific needs, emphasized materiality (focusing on the most relevant issues), and provided clearer guidance on measuring and disclosing sustainability performance.

Q4: What role do environmental pressure groups play in influencing sustainability reporting practices, as mentioned in the chapter?

A4: Environmental pressure groups play a critical role in holding corporations accountable for their environmental and social impacts. They often highlight issues such as environmental degradation, lack of transparency, and unethical practices, pushing companies to improve their sustainability reporting and practices. For example, the chapter mentions allegations against a transnational corporation (TNC) for gas leaks, insufficient disclosure, and corruption, which underscore the need for greater accountability and transparency in sustainability reporting.

Q5: How can universities and academic institutions contribute to the advancement of sustainability reporting, as discussed in the chapter?

A5: Universities and academic institutions can contribute by integrating sustainability reporting into their curricula, conducting research on best practices, and serving as role models by publishing their own sustainability reports. The chapter highlights the case of Epoka University, which aims to establish a Center of Sustainability to deepen its involvement in sustainability matters. Such initiatives can help educate future leaders, promote awareness, and drive the adoption of sustainability reporting practices across industries.

Q6: What are the potential risks for companies that fail to adopt sustainability reporting, as implied in the chapter?

A6: Companies that fail to adopt sustainability reporting risk losing stakeholder trust, facing regulatory penalties, and damaging their reputation. The chapter highlights the case of a TNC accused of environmental harm and corruption, which could lead to legal and financial consequences, as well as loss of investor and consumer confidence. Additionally, companies that ignore sustainability reporting may miss opportunities to improve operational efficiency, reduce risks, and align with global sustainability trends, putting them at a competitive disadvantage.

1.5 Summary

Chapter 1 sets the stage for understanding sustainability reporting as a vital tool for modern businesses. It highlights the importance of transparency, accountability, and stakeholder trust while acknowledging the challenges faced by smaller economies and organizations. The chapter also underscores the evolution of reporting standards and their role in improving the quality and consistency of sustainability reports. By addressing these themes, the chapter provides a comprehensive introduction to the concept and significance of sustainability reporting in today's globalized world.

Chapter 2
Founding and Objectives of GRI

Review: Discusses the establishment of GRI and its foundational goals.

Key Topics: GRI's origins, purpose, and global influence.

Outcome: Chapter 2 highlights the founding of GRI in 1997, its mission to standardize sustainability reporting, and its global influence. GRI emerged to address the need for transparency in economic, social, and environmental practices, providing frameworks like the Sustainability Reporting Guidelines (2000). Its mission is to make sustainability reporting a standard practice, enabling organizations to report impacts credibly and comparably. GRI's global influence is evident in its adoption by over 100,000 companies, fostering transparency, accountability, and sustainable development worldwide.

2.1. Introduction

History and Establishment of GRI One of the key elements for a firm to improve its sustainable and comprehensive report management is through reporting its non-financial activities. The Global Reporting Initiative (GRI) is a non-profit organization, established in Boston of the United States in 1997. It was created with the perspective that sustainability practices have an integrative role on economical, social, and environmental areas, and such practices need to be shared openly due to the common nature of sustainability challenges. GRI strongly believes that with the right tools and frameworks empowering companies to integrate equitable relationships between planetary boundaries, social equity, and economic sustainability, a truly sustainable capitalism can realize.

Mission and Vision of GRI The first GRI publication was "Sustainability Reporting Guidelines" in 2000. The publication was gathered the common grass roots efforts by the experts from 29 countries. GRI's mission is to make sustainability reporting standard practice by providing guidance and support; its goal is to allow for organizations to report on their impacts and their positive contribution to sustainable development in a credible and comparable manner. It is GRI's belief that creating a standardized system to communicate sustainable performance, companies are able to utilize their data more freely and efficiently and stakeholders are able to understand and react to such reports more pragmatically and effectively. With the belief that transparency and measurable changes can empower markets to maximize the benefits positive actions provide to society and the planet over time, GRI aims to focus on continual innovation and growth to ensure that such tools remain relevant and meaningful around the world.

2.2. History and Establishment of GRI

Review: Explores the historical context and founding of GRI.

Key Topics: Key milestones, founding organizations, and early challenges.

Outcome: Illustrates GRI's journey from inception to becoming a global leader in sustainability reporting.

GRI was founded in 1997 for associations of laundries in 20 countries that are directly affiliated with 32 multinational companies specialized in textile production in Europe. GRI received immense interest in the world when it was established, and it has been a significant development towards respecting environmental and social rights. GRI has a reporting standard regarding environmental and social issues involving organizations or businesses (Pavaloaia et al., 2017).

The Global Reporting Initiative (GRI) has become one of the most important standards for businesses and non-profit organizations to report their economic, environmental and social performance today. Finally, the conduct and contribution of the qualitative research were explained. Data derived from the content analysis of the 2016 reports published by non-profit organizations in Europe were analyzed. The research conducted in Turkey showed not to mention the report on the website of the non-profit organizations, there is no transparency in this respect. GRI provides an advantage to over 2,500 publications on the website against the report obligation of over 100,000 companies. Forty-eight out of 215 organizations reporting something as GRI were seen to provide resources in their financial structure as compared to cost advantages. There are not many studies in the literature comparing non-profit organizations by being excluded from the sample frame although they are mentioned in European countries. Turkey TUSEV, Turkey GRI also works in ensuring the dissemination of the sector companies in Turkey for a long time in GRI Turkey to make preparations and arrangements related to becoming a party the legal regulations prepared are also thought to be useful in this context.

2.3. Mission and Vision of GRI

Review: Examines GRI's mission and vision for sustainability reporting.

Key Topics: Promoting transparency, accountability, and sustainable development.

Outcome: Clarifies GRI's commitment to driving global sustainability through reporting.

As part of the promotional activities on sustainable development, in 1997 the Coalition for Environmentally Responsible Economies (CERES), the United Nations Environment Program (UNEP), and the World Business Council for Sustainable Development have created the Global Reporting Initiative (GRI). The mission of GRI is to become the world's first option in determining the sustainability reporting standards. To this end, the organizations of the international forum take steps to set up the "sustainability indicatives" in 2007, which are universal reporting standards directed at stakeholders. The promotion of environmental practices is meant to be ensured through the integration of environmental costs in the financial accounts of European business organizations (Pavaloaia et al., 2017). This financial statement is to be considered informative, not as a postulate expressing the image of economic performance or the existing financial situation.

The ways in which environmental costs are to be reported, the concepts "cost" and "environmental cost," as well as the sequencing of operations in promoting such a regulation are the main elements of debate in the financial-audit environment. The purpose of integrating the environment is to make European business organizations sensitive to a cleaner and better environment. In view of the fact that business, along with governments and the public, has an increasingly important role in this regard, economic entities are obliged to make a constructive contribution to sustainable development.

2.4 Challenging Questions and Answers

Q1: What motivated the establishment of GRI in 1997, and how has its purpose evolved since then?

A1: GRI was established to address the lack of transparency in sustainability practices, driven by the need to integrate economic, social, and environmental dimensions into business reporting. Its purpose has evolved from creating basic guidelines to developing comprehensive frameworks like the GRI Standards, which are now globally recognized for enabling credible and comparable sustainability reporting.

Q2: How does GRI's mission align with the United Nations Sustainable Development Goals (SDGs)?

A2: GRI's mission aligns with the SDGs by promoting transparency and accountability in sustainability reporting. Its frameworks help organizations measure and communicate their contributions to global goals like climate action, gender equality, and responsible consumption, driving collective progress toward sustainable development.

Q3: What challenges did GRI face in gaining global acceptance, and how were they addressed?

A3: GRI faced challenges such as skepticism about voluntary reporting and the complexity of creating universal standards. It addressed these by collaborating with international organizations like CERES and UNEP, engaging stakeholders, and continuously refining its frameworks to ensure relevance and practicality.

Q4: How does GRI ensure its standards remain relevant in a rapidly changing sustainability landscape?

A4: GRI ensures relevance by focusing on continual innovation, updating its standards to address emerging issues like climate change and human rights, and engaging with stakeholders to adapt frameworks to evolving global sustainability challenges.

Q5: What role does GRI play in promoting transparency and accountability in non-financial reporting?

A5: GRI promotes transparency by providing standardized frameworks for reporting non-financial impacts, enabling organizations to disclose their sustainability performance in a credible and comparable manner. This fosters accountability and builds trust among stakeholders, including investors, customers, and communities.

2.5 Summary

Chapter 2 explores the founding of GRI in 1997, its mission to standardize sustainability reporting, and its global influence. It highlights GRI's evolution, frameworks, and commitment to transparency, accountability, and sustainable development, positioning it as a leader in shaping sustainable business practices worldwide.

Chapter 3

GRI Reporting Frameworks and Guidelines

Review: Provides an overview of GRI's frameworks and guidelines.

Key Topics: Structure of GRI Standards, reporting principles, and application.

Outcome: Chapter 3 explores GRI's reporting frameworks and guidelines, emphasizing their role in standardizing sustainability reporting. It introduces the GRI Standards, including Universal and topic-specific standards, and explains the Core and Comprehensive options for reporting. The chapter highlights how these tools provide flexibility and comprehensiveness, enabling organizations to tailor reports to their needs while ensuring transparency and accountability. It also underscores the importance of materiality and stakeholder engagement in creating credible and impactful sustainability reports.

3.1. Introduction

When the Global Reporting Initiative (GRI) was founded in 1997—only 21 years ago—the expectation of more sustainable corporate behaviour seemed almost naive. Today, the GRI certificate is of utmost importance for the recording of sustainability-relevant data in every third company in Germany. Sixty-two percent of the middle and large German companies are based on the GRI guidelines.

There are versions with content for various groups of recipients, also for the creative sector and small and mid-size enterprises. Guidelines for very demanding sustainability analysis are provided. Thus, conditions for the arithmetic calculations of impacts have been established. The GRI certificate is to emphasize the credibility of the information and the ability of GRI processing to analyze the sustainability performance of every type of organization experienced so far.

Public commitments should be easier through transparency (Calitz & Zietsman, 2018). The GRI reporting formats are based on the GRI Reporting Framework for updating it to the most recent versions. There are GRI 2002, GRI 2006, GRI G2, G3, G3.1, and now G4, but only the most recent G3.1 and some of the new G4 principles are provided. EN exceeds this restriction.

For the first report starting 2014, the GRI "Core" option must comprise information on 33 general standard disclosures; 16 from them [9 Indicators, 3 Aspect Boundaries, Aspects, and Impacts modeled; 2 from the Management Approach]; 4 from these [6 Unit of measure, Data Information 1–4]. The expected content of the Sustainability Report is given by the GRI principles who strictly refer to the Guidelines. Thus, the issues are raised as treated in the GRI Principles. An expectation of a transparent, honest report on sustainability, despite the GRI certificate for fantastic analytic capacity, is also given. On this subject, a careful arrangement of content and its deep disclosure is suggested. Some further instructions are developed that meet the need for them after the negligent review of hundreds of analyst reports. Everything here is "text2". The Creative Sector Supplement was not considered from the outset for English-language analysis and has not even been analyzed para-linguistically. In-depthness is also foregone on that, due to the time framework, SIC classification items, and the immeasurable valuation of creative procedures to derive what the issues are.

3.2. GRI Standards Overview

Review: Introduces the GRI Standards and their components.

Key Topics: Universal Standards, topic-specific standards, and modular structure.

Outcome: Highlights the flexibility and comprehensiveness of GRI Standards for diverse organizations.

Founded in 1997, the Global Reporting Initiative (GRI) is a network-based organization known as pioneer to create the first framework for sustainability reporting. In 2000, as the GRI Guidelines launched, UN Conference on Trade and Development invited and first involved 50 world's largest corporations to eliminate corruption, promote sustainability and respect for human rights. The formation of GRI marks the birth of the sustainability issues as the priority on political and corporate agendas. Ever since the inception of the GRI guidelines, organizations worldwide have felt an increasing pressure to report on their CSR activities and to monitor them. However, the adoption of the guidelines is not being followed equally within countries or industries. Furthermore, little academic research yet exists that observes the industry-specific proneness of sustainability reporting (Wuerzinger, 2015). This is a peculiar area since some of the detrimental side effects of reporting are already visible or being told, such as that it can redirect the managers to engage in legitimacy-driven, 'cheap talk' activities instead of building the actual corporate sustainability. In April, 2018, the GRI announced that G4 Guidelines translation in ten languages was omitting the specific standard disclosures and that they would not publish or offer any service any longer on these translations. They also declared that version 3.1 of the Guidelines are no longer available; however, it is still available in the GRI Database. Nevertheless, the absence of these translations on the main GRI database may lead to the predicament of how, especially in the EU, the largest and most diverse trading block in the world, companies would now report according to GRI.

3.3. Core and Comprehensive Option Frameworks

Review: Explains the differences between Core and Comprehensive options.

Key Topics: Reporting levels, disclosure requirements, and organizational needs.

Outcome: Helps organizations choose the appropriate reporting level based on their goals and resources.

The international standards, ISO 26000 and the Global Reporting Initiative (GRI), were launched in conjunction with the proposal to help firms in their implementation of CSR. The GRI guidelines include general disclosures on several aspects of the business (aspects, management approach, report scope and basis of preparation) and specific standard disclosures (Aspects and Indicator Protocols). The last part of the specific standard disclosures are therefore not found in the GRI guidelines, but contain an alphabetical index and protocols (Wuerzinger, 2015). Anyways the new guidelines, much hope this will cause more attention on the materiality principle. According to GRI for this a mapping of the most influential economic, environmental and social (labour practices, decent work, human rights, society, product responsibility) indicators based on GRI G4 are required and it is necessary to determine the most material sustainability topics that fall into Boundary 3 areas, including the under different types of material topics.

In addition, most organizations report additional details on their approach to stakeholder engagement, strategy and analysis, governance and the identification and management of risk, where formalised or implemented on an ad hoc basis; therefore a materiality mapping was considered for these subject otherwise this analysis is limited to the materiality of the report on the systemic strategies to identify material aspects, and does not extend to indicators related to selected sectors of the economy nor to indicators related to the monitoring of impacts along the value chain. Further, of the 91 Specific Standard Disclosures, G4 provides for different segments of the economy to report on different aspects. Companies thus need to determine all the specific standard disclosures that are to be reported for the organisation as a whole. Based on statements by Lothar Rieth of the GRI organization, starting in the 2015 fiscal year, organizations are expected to disclose specific standard disclosure on material aspects, regardless of the claimed sector of the economy. This suggests a change of policy over time; in the future, only specific standard disclosures related to identified material aspects may have to be disclosed (Romana, 2018). GRI will not provide guidelines from 2015 onwards and will not offer services for these guidelines anymore, encouraging organizations to move towards their new guidelines. Note that the 100 applications were received shortly before the release of the 'G4 Content Index Template Table' and have been decided to not provide the specific content of the specific standard disclosures, which was often requested as well.

3.4 Challenging Questions and Answers

Q1: What are the key differences between the Core and Comprehensive options in GRI reporting?

A1: The Core option requires fewer disclosures and is suitable for organizations starting their sustainability reporting journey. The Comprehensive option demands more detailed disclosures, catering to organizations aiming for higher transparency and accountability. Both options allow flexibility based on organizational goals and resources.

Q2: How do GRI Standards ensure materiality in sustainability reporting?

A2: GRI Standards require organizations to identify and prioritize material topics—those most relevant to their business and stakeholders. This ensures that reports focus on significant economic, environmental, and social impacts, enhancing relevance and credibility.

Q3: What challenges do organizations face when adopting GRI frameworks?

A3: Challenges include determining material topics, aligning reporting with organizational goals, and managing resource constraints. Additionally, transitioning to newer GRI versions, like G4, can be complex due to changes in disclosure requirements.

Q4: How does GRI's modular structure benefit organizations?

A4: The modular structure allows organizations to select relevant standards (Universal or topic-specific) based on their industry and sustainability priorities. This flexibility ensures that reports are tailored, comprehensive, and aligned with global best practices.

Q5: Why is stakeholder engagement critical in GRI reporting?

A5: Stakeholder engagement ensures that reports address the concerns and interests of key stakeholders, enhancing transparency and trust. It also helps organizations identify material topics and improve the quality of their sustainability disclosures.

Q6: How has GRI evolved to address emerging sustainability challenges?

A6: GRI has continuously updated its frameworks, such as transitioning from G3.1 to G4, to address emerging issues like climate change and human rights. This evolution ensures that its standards remain relevant and effective in promoting sustainable development.

3.5 Summary

Chapter 3 delves into GRI's reporting frameworks, including Universal and topic-specific standards, and the Core and Comprehensive options. It emphasizes materiality, stakeholder engagement, and flexibility in reporting, enabling organizations to create credible, tailored sustainability reports. The chapter highlights GRI's role in promoting transparency and accountability in sustainability practices globally.

Chapter 4
Benefits of GRI Reporting

Review: Discusses the advantages of adopting GRI reporting for organizations and stakeholders.

Key Topics: Improved reputation, risk management, and stakeholder trust.

Outcome: Chapter 4 highlights the dual benefits of GRI reporting for organizations and stakeholders. For organizations, GRI reporting enhances decision-making, operational efficiency, and competitive advantage by providing standardized metrics for sustainability performance. For stakeholders, it fosters transparency, accountability, and informed decision-making, building trust and engagement. The chapter underscores GRI's role in reducing information asymmetry and promoting global sustainability standards, making it a vital tool for both internal and external value creation.

4.1. Introduction

Many companies have recognized the importance of sustainability issues and began to disclose sustainability performances, either in stand-alone reports or as part of an annual integrated report (Pavaloaia et al., 2017). After reviewing a sample of 113 listed manufacturing companies in Taiwan over 2003–2011, found that the adoption of GRI reporting becomes more prevalent among larger firms and firms with concentrated ownership structures. The results suggest that GRI is one of the most important communication tools regarding environmental issues for reducing the level of asymmetric information between organizations and their stakeholders.

4.2. Organizational Benefits

Review: Focuses on the internal benefits for organizations using GRI.

Key Topics: Enhanced decision-making, operational efficiency, and competitive advantage.

Outcome: Shows how GRI reporting drives organizational performance and sustainability.

GRI representatives underline that the objective of GRI is to promote sustainability beyond the organization itself, also considering the influence of the organization's activity on the "sustainability trio": 1. Environment, social, financial 2. Region and country in which it operates 3. its influence globally - a significant competitive advantage is that with GRI standards, the metrics of each economy are compared, considering the fact that currently the metrics are established at national level, are different and have to wind up different requirements of information. Benefiting from the active participation of representatives of businesses, multinational corporations, non-profit organizations, governments, academic, union and civil society organizations, and more than 4000 persons attending the public debates on the G4 version and mandatory BSE reporting, the GRI standards are revised periodically, in principle every five years, following the three-year development process. Through a series of public debates and comments on exposure drafts, stakeholders are involved in revising and achieving an improvement over previous regulations, offering an easier use and more relevant, comparable and verifiable indicators.

GRI standards consist of 36 indicators, grouped into two categories: 3 foundation and 33 specific indicators, organized according to 34 aspects related to the organization's impact on the economy, the environment or society and the global compact principles. In addition to the "general standard protocol" and the "management approach" to each indicator, performance is measured using descriptive information on disclosures, progress on commitment targets, and related standards and parameters. Being a voluntary reporting, the revision of GRI has included levels of certifications that reflect the extent to which the GRI standards were applied. It is not expected to implement an organization moving from one level of certification to another, but it has the option to choose the option that best suits its reporting needs and the information needs of the stakeholders.

4.3. Stakeholder Benefits

Review: Explores how stakeholders benefit from GRI reporting.

Key Topics: Transparency, accountability, and informed decision-making.

Outcome: Highlights the role of GRI in building trust and engagement with stakeholders.

The interest in communication of the economic, social and environmental effects pursues by sales growth, cost reduction, risk management, human resources, establishing long-term relationships with stakeholders and ensuring economic sustainability. Corporate reporting on sustainable development through the databases of the Global Reporting Initiative (GRI) is often cited as the most influential and used system at international level (Pavaloaia et al., 2017). The literature on the impact of reporting on sustainable development according to GRI on the organization and its relations with stakeholders is scarce.

Although companies were the first organizations to report on sustainable development using GRI, the annual performance reports created respect for the European Commission's recommendation, which instead focused on public bodies. The database includes 15 indicators, or rather socio-economic and environmental objectives regarding European policies: innovation and human resources, the single market, social cohesion and employment, protection and human health, animal health and the environment. Goal results aimed at improving the quality of air represent the highest percentage of achievements, and those regarding the Lisbon strategy have the fewest results. It is shown that local authorities, representing the most numerous samples, lag behind expectations. On the other hand, the European Committee, which launched the initiative, is over-exceeded in the achievement of five of the six objectives published.

4.4. Challenging Questions and Answers

Q1: How does GRI reporting contribute to reducing asymmetric information between organizations and stakeholders?

A1: GRI reporting provides standardized, transparent, and comparable metrics on sustainability performance, ensuring stakeholders have access to reliable information about an organization's environmental, social, and financial impacts.

Q2: What are the three pillars of the "sustainability trio" emphasized by GRI?

A2: The sustainability trio includes environmental, social, and financial impacts, which GRI reporting addresses to promote holistic sustainability.

Q3: Why is GRI reporting particularly beneficial for larger firms and those with concentrated ownership structures?

A3: Larger firms and those with concentrated ownership often face greater scrutiny from stakeholders, making GRI reporting a valuable tool for demonstrating accountability and enhancing reputation.

Q4: How does GRI reporting improve operational efficiency for organizations?

A4: By identifying areas for improvement in sustainability practices, GRI reporting helps organizations streamline processes, reduce waste, and optimize resource use.

Q5: What role do stakeholders play in the revision of GRI standards?

A5: Stakeholders participate in public debates and provide feedback on exposure drafts, ensuring GRI standards remain relevant, comparable, and user-friendly.

Q6: What challenges might local authorities face in achieving GRI-aligned sustainability goals?

A6: Local authorities often lag due to limited resources, lack of expertise, or competing priorities, making it harder to meet GRI-aligned sustainability objectives.

4.5. Short Summary:

Chapter 4 explores the benefits of GRI reporting for organizations and stakeholders. It highlights how GRI enhances organizational performance through improved decision-making, operational efficiency, and competitive advantage while fostering stakeholder trust via transparency and accountability. The chapter emphasizes GRI's role in standardizing sustainability metrics, reducing information asymmetry, and promoting global sustainability standards. It also discusses the challenges and achievements of GRI reporting, particularly in larger firms and public bodies, underscoring its value as a communication and performance improvement tool.

Chapter 5
GRI Reporting Process

Review: Outlines the steps involved in implementing GRI reporting.

Key Topics: Planning, data collection, and reporting cycles.

Outcome: Chapter 5 provides a comprehensive roadmap for organizations to implement GRI reporting effectively. It outlines key steps, including stakeholder identification, materiality assessment, and report preparation, ensuring that organizations can create relevant, transparent, and impactful sustainability reports. The chapter emphasizes the importance of aligning reporting with GRI standards to enhance comparability, consistency, and accountability. It also highlights the challenges, particularly for small and medium enterprises, and underscores the need for standardized, legally binding reporting frameworks to ensure widespread adoption and reliability.

5.1 Introduction

The first step in GRI reporting is the definition of the content of the report aimed at the relevance of the material issues. The scope of the report refers to the organization to prepare a report for its entire activity or for a certain activity segment, determining the sector in which it operates, the coverage is monitoring the subsidiary activities, joint ventures, etc., which support the fulfilment of the main activities and organizational changes or reporting periods. The boundary eliminates the most important stages of the organization's value chain, describing the preparation (in terms of selection, evaluation and control) of the items or risks inherent to the territory in the situation of the activities of the geographic profile. In 2016 over three years of research, over one million comments were submitted and the GRI standards have been created. These relate to the first four levels of the former and have been divided into three separate standards on reporting principles (applicable to any level), defining standard disclosures (general and sector-specific) and contemporary elements (aspects and key topics: economy, social and environmental impact, social relational practices, and social results) (Pavaloaia et al., 2017). As with the former, the GRI standards are intended to be used to prepare but also to better understand a sustainability report, to provide recommendations for good practice and to promote report reliability. To apply it in a coherent, complete and comparable manner, other documents could be considered where the GRI standards are supplemented and specified, as additional information is provided for developing indicators, including the topics that are common to a set of standards. The GRI standards are reported alongside the EU Directive on Financial Reporting, within a model translation for companies to prepare concise and consistent reports, and with the EC guidelines as well.

5.2. Steps to Implementing GRI Reporting

Review: Details the practical steps for adopting GRI reporting.

Key Topics: Stakeholder identification, materiality assessment, and reporting preparation.

Outcome: Guides organizations through the process of creating a GRI-compliant report.

The implementation of the Global Reporting Initiative (GRI) framework in sustainability reporting includes several steps to be considered by enterprises. The first step is to commence the implementation of the GRI with a decision on which indicator categories, or in other words, which of the Economic, Environmental, Labor and Work Conditions, Social, Human Rights, or Product Responsibility, will be chosen for disclosure in the company report. The most crucial part of the GRI guidelines for the 'what matters' analysis is the selection of sub-categories to be disclosed in the content of the chosen few predefined principle categories. The distinction is made between general and specific standard disclosures. General disclosure requires companies to give information on the selected sub-categories of the GRI guidelines to the 'fullest extent possible.' Special emphasis of this analysis is on 58 specific standard disclosures of G4 Guidelines, with the corresponding classifying sub-category letter and number. The law of the Transparency Directive requires the publication of corporate responsibility (sustainability) reports in order to enhance integration on a pan-European level. With the aim of facilitating behavior change, the reports must advise company impacts on employees, the social and environmental context, human rights, and the fight against corruption. Transparency results in accountability for a corporation's actions. However, there is disagreement in the academic literature on what transparency exactly means. GRI is the most widely used organizational framework for sustainability reporting. Corporate transparency can be achieved through GRI sustainability reporting. GRI provides advice on how to develop the content of company reports and guidance on what information should be included in them. Indicator-wise it has 87 standard explanatory disclosures and numerous sector-specific supplements. To boost the comparability, consistency, and quality of this structure, both GRI and the ECRI trade association provide training courses and online materials. Referring to the sustainability report in the printed company report remains a voluntary action. Rapid up-take of GRI reporting guidelines is unlikely. Medium and small enterprises are likely to bear high costs. Only large firms will consider the implementation of the GRI Guidelines. Such widespread divergence in approach would undermine the purported comprehensive performance to facilitate the company assessment. On that account, the EC needs to apply legally binding rules, not only examining the disclosure of behavior by company but establishing uniform reporting standards that should be followed in all reports (Ghoogassian, 2015).

5.3. Materiality Assessment in GRI Reporting

Review: Explains the importance of materiality in GRI reporting.

Key Topics: Identifying key issues, stakeholder input, and prioritization.

Outcome: Emphasizes the role of materiality in creating relevant and impactful reports.

The role of the Global Reporting Initiative (GRI) as the promoter of sustainability reporting is unquestionable, as environmental, social and economic aspects of a business need to be transparent. The main idea behind the reporting is to provide all information necessary to assess the actual impacts of a business. While transparency is regarded as an essential part of the global economic landscape (in the regulation market and reputation framework a changing perspective towards sustainability has arrived in business), and published sustainability reports are still heavily criticized in terms of marketing tools, (Wuerzinger, 2015) argues that sustainability reporting should be approached as a homotopic system with organizations developing sustainable strategies in order to comply with the legal framework, guidelines and regulations, while critics should ensure validity of the data, checking and comparing it with actual reports, previous year data and real situations. The GRI CSR reporting system consists of Standart Disclosures and indicators from the G4 Guidelines, and the Reporting principles and Standard Disclosures from the G3.1 Guidelines. The guidelines intended to represent the best practice of sustainability reporting, and the presenter of this report attempts to illustrate the possibility of integrating both versions of GRI Guidelines in a harmonized presentation. The 5.3 Analysis of the CSR reports from 2008 to 2013, comprising the mapping of the materiality matrices present in the report.

5.4. Challenging Questions and Answers

Q1: What is the first step in the GRI reporting process, and why is it important?

A1: The first step is defining the report's content and scope, focusing on material issues. This ensures the report addresses the most relevant sustainability impacts, aligning with stakeholder expectations and organizational goals.

Q2: How does materiality assessment contribute to the effectiveness of GRI reporting?

A2: Materiality assessment identifies key issues that matter most to stakeholders and the organization, ensuring the report is relevant, focused, and impactful.

Q3: What challenges do small and medium enterprises (SMEs) face in adopting GRI reporting?

A3: SMEs often face high implementation costs and resource constraints, making it difficult to comply with GRI standards compared to larger firms.

Q4: How do GRI standards enhance the comparability and consistency of sustainability reports?

A4: GRI provides standardized metrics and guidelines, ensuring reports are consistent, comparable, and reliable across organizations and industries.

Q5: What role do stakeholders play in the GRI reporting process?

A5: Stakeholders provide input during materiality assessments and help prioritize key issues, ensuring the report reflects their concerns and enhances transparency.

Q6: Why is transparency critical in GRI reporting, and how does it benefit organizations?

A6: Transparency builds stakeholder trust, enhances accountability, and improves organizational reputation by demonstrating a commitment to sustainability and ethical practices.

5.5. Summary

Chapter 5 outlines the GRI reporting process, detailing steps like stakeholder identification, materiality assessment, and report preparation. It emphasizes the importance of aligning reports with GRI standards to ensure relevance, transparency, and comparability. The chapter highlights challenges, especially for SMEs, and calls for standardized reporting frameworks to promote widespread adoption. By providing a clear roadmap, it enables organizations to create impactful sustainability reports that meet stakeholder expectations and enhance accountability.

Chapter 6

GRI and Sustainable Development Goals (SDGs)

Review: Examines the alignment between GRI and the UN SDGs.

Key Topics: Mapping GRI indicators to SDGs, contribution to global goals.

Outcome: Chapter 6 demonstrates how GRI reporting aligns with and supports the United Nations Sustainable Development Goals (SDGs). It highlights the complementary nature of GRI standards and SDG frameworks, emphasizing shared objectives like transparency, accountability, and global sustainability. The chapter shows how GRI indicators map to SDGs, enabling organizations to track and report their contributions to global goals. It also underscores GRI's role in enhancing SDG reporting through standardized metrics, fostering collaboration between public and private sectors, and promoting sustainable development worldwide.

6.1 Introduction

The GRI guidelines consist of economic, ecological and societal aspects. In 2014, the GRI adjusted and extended the guidelines, whereas the Corporate Sustainability and Reporting Directive in the EU sets an obligation for CSR reporting including a unique identifier as the guidelines are so recommendation to be applied for such reporting. As all industries are affected by the guidelines, sustainability initiatives are intended to be increased within the food industry. In order to measure the industry's possible sustainability improvement, environmental, social and economic data need to be analyzed. Thus, the purpose of the previous project was to examine the industry's present state in terms of sustainability according to the GRI guidelines. The examined data should be helpful since it contains a lot of information that is mandatory to be disclosed according to the GRI. On the other hand for the purpose of the project, the risk arises that the companies could submit false or biased data, which makes the validation of their sustainability more difficult.

In 2016, the United Nations states the sustainable development goals of the 2030 Agenda for Sustainable Development as "an action plan for people, planet and prosperity". Therefore, these goals contain new and innovative aspects, whereas they do not refer only to the industry but take a global approach. It is crucial and necessary for EU businesses to understand and specify the industry's contribution to the SDGs. That is why the GRI prompts its sector supplement for the food industry. In this context, the performance of the food industry companies should be evaluated during a time period and a potential voluntary GRI reporting of a significant number of these could be assumed.

6.2. Alignment of GRI with SDGs

Review: Explores how GRI Standards align with the SDGs.

Key Topics: Shared objectives, reporting synergies, and global impact.

Outcome: Demonstrates the complementary nature of GRI and SDG frameworks.

The Global Reporting Initiative (GRI) sustainability reporting standards (SRS) were established to enhance the organization's transparency and accountability toward sustainability issues. This action was initiated to promote the sustainability of different sectors of organization such as economic, environmental and social. Sustainability development goal (SDG) is an inclusive agenda set by United Nations to be attained by year 2030 for the wellbeing of future generation and the ecosystem. To deliver this goal, inclusion of GRI is crucial and compatible.

In the SDG compliance discourse, both private and public sectors are actively engaged. The active involvement aimed to address the underlying primary challenge; to strengthen the indicator's deployment pertaining to SDGs. The underemployment of indicator may also relate to the misspecification in the government unveiling of inclusive policies, which falls under the category of policy trap hypothesis. A manifestation of both policy trap hypothesis and loose condition attached to policy instruments were seen in the banks (Avrampou et al., 2019).

6.3. Contribution to SDG Reporting

Review: Discusses how GRI reporting aids in SDG progress tracking.

Key Topics: Metrics, accountability, and transparency.

Outcome: Highlights GRI's role in advancing SDG reporting and accountability.

Boosting the contribution to the achievement of the Sustainable Development Goals of the United Nations is one of the current actions promoted by the GRI to the adopters of its model. To this end, it has developed a project in collaboration with the acronym (S1T2A3R4S) to collect in a database all the contributions of each organization to the 17 SDGs through the dissemination of informative social accounts. In addition, within its set of social indicators, new calculations have been included that allow measuring the concrete work of a society on the achievement of the SDGs, both as an ecosystem organization with the adoption of the Global 2030 Agenda, and from the functional work as an institution, in areas such as social action, collaboration with the third sector, etc. The GRI model so far is internationally the most widely used model for sustainability, offering the most advanced construction model for social accounting (Barkemeyer et al., 2015). At the beginning of this chapter, and since the European Regional Development Fund is the manager of the GRI database of Spanish organizations of the ERDF, an analysis of the descriptive parameters of the set of social accounts of these entities, based on the information offered by the GRISET database.

The achievement of the Sustainable Development Goals becomes a major challenge to territorial organizations and companies that carry out their activity in them. It involves deeply linking decisions regarding business models, areas of territorial and sectoral influence, and also regarding collaboration with the different Affairs in which the organizations intervene in the different instances of their management. The SDGs and their implementation model (2030 Agenda) are world objectives of a very diverse nature that include equality, water, work, energy, climate, education, and health, among others, with common values such as stewardship of the planet, peace and prosperity for individuals and societies, just, equitable and rule environment, and all through a broad-based collaboration between the public and agents.

6.4. Challenging Questions and Answers

Q1: How do GRI standards align with the UN SDGs?

A1: GRI standards provide a framework for reporting on economic, environmental, and social impacts, which directly align with the SDGs' focus on global sustainability and development.

Q2: Why is GRI reporting crucial for tracking progress toward the SDGs?

A2: GRI reporting offers standardized metrics and transparency, enabling organizations to measure and communicate their contributions to the SDGs effectively.

Q3: What challenges arise in aligning GRI indicators with SDG reporting?

A3: Challenges include ensuring data accuracy, avoiding biased reporting, and addressing the complexity of mapping GRI indicators to specific SDG targets.

Q4: How does GRI reporting enhance accountability in achieving the SDGs?

A4: By providing transparent and comparable data, GRI reporting holds organizations accountable for their sustainability efforts and progress toward SDG targets.

Q5: What role do public and private sectors play in advancing SDG reporting through GRI?

A5: Both sectors collaborate to strengthen SDG indicators, address policy gaps, and promote sustainable practices, ensuring a unified approach to global development.

Q6: How does GRI support the food industry in contributing to the SDGs?

A6: GRI provides sector-specific supplements, like the food industry supplement, to help companies measure and report their sustainability performance, aligning with SDG objectives.

6.5. Short Summary:

Chapter 6 explores the alignment between GRI reporting and the UN SDGs, emphasizing their shared goals of transparency, accountability, and sustainability. It highlights how GRI indicators map to SDGs, enabling organizations to track and report their contributions to global development. The chapter underscores GRI's role in enhancing SDG reporting through standardized metrics and fostering collaboration between public and private sectors. By integrating GRI and SDG frameworks, organizations can effectively measure and communicate their impact, advancing sustainable development worldwide.

Chapter 7
GRI and Stakeholder Engagement

Review: Focuses on the importance of stakeholder engagement in GRI reporting.

Key Topics: Dialogue, feedback, and inclusivity.

Outcome: Chapter 7 underscores the critical role of stakeholder engagement in GRI reporting, emphasizing its importance for building trust, ensuring relevance, and enhancing accountability. By involving stakeholders in the reporting process, organizations can create more meaningful and credible sustainability reports. The chapter highlights how effective engagement fosters dialogue, inclusivity, and feedback, ultimately improving report quality and aligning corporate practices with stakeholder expectations. It also provides best practices for engaging stakeholders, offering practical strategies for organizations to enhance their sustainability reporting efforts.

7.1 Introduction

Climate change, human rights violations, natural disasters and scarcity of natural resources have, for the last two decades, contributed to a transformation in the way companies are welcomed to conduct their business. As the spotlight has been directed towards pressures resulting from a heavily industrialized society, the general public and stakeholders have seen both the social and environmental impact of corporate activities. As a result of this increased general awareness and the willingness of businesses to maintain standards and actively work towards creating shared value, stakeholders, such as state supervisory authorities, non-governmental organizations, employees and the natural environment, as well as the companies suppliers and clients, have demanded more information and created pressure on how companies perform on sustainability (Arvidsson, 2018). Among the different means that companies use in order to communicate their relationships with stakeholders, reports, especially sustainability ones, still are the objects of intense academic research, being widely viewed as a key mean to report how they make progress responding to increasing society pressures for a more responsible and sustainable conduct. Since the beginning, the prime focus of the business and the stakeholders has become the plainly economic performance of a company. In the area of the quality of life of a particular distribution network, criminal behavior corruption and the normal business operations. Businesses and stakeholders held together engagement work as a mechanism for generating company success performance which is crucial to the success of the company and is contributing to value creation for shareholder distributors and large classes of stakeholders. At a unique business level, engagement returns have a telecommunications company in the hard market case with a potential contribution of interaction with key stakeholders possible with the greatest reputational, economic or strategic impacts. At a more refined market level, engagement activity can favor the development of nation-wide a shared communications plan between key stakeholders, such as regulatory authorities and public administrations, with a specific focus on full customer satisfaction and customer satisfaction quality levels, and ability improvement.

7.2. Importance of Stakeholder Engagement in GRI Reporting

Review: Explains why stakeholder engagement is critical for GRI reporting.

Key Topics: Trust-building, relevance, and accountability.

Outcome: Shows how engagement enhances report quality and credibility.

This research analyses the effect of company attributes and country-level determinants on the stakeholder engagement levels of companies in regard to reporting practices in accordance with the Global Reporting Initiative (GRI) guidelines. It is argued that the creation and annual update by companies of this specific performance record could increase the intensity of interactions that potentially have adverse as well as supportive effects on the choices of individual companies regarding the content and form of their GRI reports (Barkemeyer et al., 2015). Beyond macro-level observations of a global increase in the number of sustainability reports, company-level evidence analyzing the economic determinants of corporate sustainability reporting and its quality and impact is still rather limited. So far, most studies on sustainability reporting based on GRI indicators have either focused on content analysis or have been stakeholder-oriented and thus pursued a normative rather than a positive research interest. There is no research considering the effect of company characteristics. Similarly, CSR reports of Chinese companies should be critically investigated.

Notification of corporate performance in regard to a systematic integrated materiality-focused sustainability framework could potentially be of vital importance for the resolution of sustainability problems, and the GRI's philosophy and methodology are held to embody these aspirations during the last years. In fact, other research has found clear evidence of the growing aspirations of developing country firms to become involved in the shaping of the transnational norms, principles and regulative capacities. Using China as a case, this research delves into the involvement of companies in GRI reporting practices, and evaluates the 'effectiveness' of the GRI regulatory system in terms of output, outcome and impact of governance effectiveness. An understanding of the outcomes of this study may be of interest to environmental groups and other stakeholders willing to engage more effectively with the development of transnational systems such as the GRI. For audit purposes, the GRI also relies on cross-verification and stakeholder dialogue; however, the success of these mechanisms has hardly been researched. In this instance, our understanding of the GRI still severely falls short of a full assessment.

7.3. Best Practices for Engaging Stakeholders

Review: Provides strategies for effective stakeholder engagement.

Key Topics: Communication, inclusivity, and feedback mechanisms.

Outcome: Offers practical guidance for organizations to engage stakeholders effectively.

Given the radically new landscape of global politics and trade, with profound implications for the political and regulatory contexts in which companies operate, and an increased complexity and intensity of stakeholder demand, which in particular reflects the politico-regulatory pressures that are placed on firms, they considered the reporting of socio-environmental information. While still marginal in comparison to financial reporting, a steadily growing number of companies choose to report their social and environmental performance.

Regarding the likelihood of reporting, while some companies headquartered in the Global North, and specifically those from Australia, are somewhat more likely than the global median to publish a sustainability report, most firms, and in some countries even this basic level of compliance with standards, cannot be taken for granted. With regard to report content, an examination of a diverse array of developed and developing countries reveals that there are country-level differences in the reporting of social and environmental information. Specifically, companies headquartered in the Global North overall pay more attention to some socio-environmental issues than their peers from the Global South. These issues specifically include certain Eco-Efficiency Indicators developed by the standards, but encompass other environmental indicators too. Moreover, companies from the Global North also place a more defined emphasis on workplace-related social issues than on the social issues that other indicators refer to.

7.4. Challenging Questions and Answers

Q1: Why is stakeholder engagement considered essential for GRI reporting?

A1: Stakeholder engagement ensures reports are relevant, credible, and aligned with stakeholder expectations, fostering trust and accountability.

Q2: How does stakeholder engagement improve the quality of sustainability reports?

A2: By incorporating stakeholder feedback, reports become more comprehensive, accurate, and reflective of real-world impacts, enhancing their overall quality.

Q3: What challenges do organizations face in engaging stakeholders effectively?

A3: Challenges include managing diverse stakeholder interests, ensuring inclusivity, and maintaining consistent communication throughout the reporting process.

Q4: How does stakeholder engagement contribute to corporate accountability?

A4: Engaging stakeholders holds organizations accountable for their actions, ensuring transparency and encouraging responsible business practices.

Q5: What are some best practices for effective stakeholder engagement in GRI reporting?

A5: Best practices include clear communication, inclusive dialogue, regular feedback mechanisms, and integrating stakeholder input into decision-making processes.

Q6: How does stakeholder engagement differ between companies in the Global North and Global South?

A6: Companies in the Global North often prioritize workplace-related social issues and eco-efficiency, while those in the Global South may focus on broader socio-environmental challenges, reflecting regional priorities.

7.5. Summary

Chapter 7 highlights the importance of stakeholder engagement in GRI reporting, emphasizing its role in building trust, ensuring relevance, and enhancing accountability. It explains how involving stakeholders in the reporting process leads to more meaningful and credible sustainability reports. The chapter provides best practices for effective engagement, such as fostering dialogue, inclusivity, and feedback mechanisms. It also explores regional differences in stakeholder engagement, noting that companies in the Global North and South prioritize different socio-environmental issues. Overall, the chapter underscores the value of stakeholder input in creating impactful and transparent sustainability reports.

Chapter 8
GRI and Assurance

Review: Discusses the role of assurance in GRI reporting.

Key Topics: Types of assurance, credibility, and challenges.

Outcome: The chapter emphasizes the critical role of assurance in enhancing the credibility and reliability of GRI sustainability reports. It explores various assurance methods, their benefits, and challenges, highlighting how external assurance can improve transparency and stakeholder trust. However, it also notes the complexities, such as cost and resource requirements, and questions the direct impact of assurance on the overall quality of disclosed information.

8.1 Introduction

Transnational forms of private governance are increasingly used to address sustainability challenges. Corporate sustainability reporting according to the Global Reporting Initiative (GRI) is a transnationally governed self-regulation regime that aims to enhance the transparency and credibility of corporate sustainability management. Private transnational governance regimes such as the GRI are considered additional to public governance solutions (Barkemeyer et al., 2015). In consequence, they can be used to narrow governance gaps connected to formal public legislation, to encourage progressive 'beyond-compliance' behavior, or to otherwise regulate new types of non-financial behavior. The effectiveness of transnational governance regimes comparing the situation faced by the regime to the situation targeted by this regime, and assessing if the governance regime has been able to close the identified governance gap. The effectiveness of the GRI regime regarding a set of pre-determined effectiveness criteria in up to 40 countries.

Several thousand companies based in up to 40 countries have been observed in terms of their sustainability reporting over a period of at least five years. Content analysis was used to longitudinally compare the characteristics of nearly 7,500 sustainability reports from up to these 40 countries. Specifically, whether there is evidence of an overall shift in the quality and content of reports over time, as suggested by the rhetoric of the GRI Secretariat and some of its other promoters. The number of reports based on GRI guidelines is assessed, and to what extent reporting according to the GRI (GRI-plus) differs from other types of (GRI-based) reporting in terms of the report's content and quality.

8.2. Types of Assurance in GRI Reporting

Review: Explores different assurance approaches for GRI reports.

Key Topics: Internal vs. external assurance, levels of assurance.

Outcome: Helps organizations choose the appropriate assurance method.

With the lecturing of the implementation of the Global Reporting Initiative framework in partial fulfilment for the degree of Master of Commerce in Accounting and Finance. Preparation and assurance of sustainability reports has experienced high growth over the past decade. This reflection is part of an ongoing examination of sustainability reporting developments in Australia.

GRI has been utilized as a benchmark in studies examining sustainability reporting trends. An empirical analysis using annual reports has shown that GRI guidelines have been a significant reference in sustainability reporting practice. However, this reporting behaviour is not mandatory due to the absence of any regulation. Although a meta-analysis study did not identify a significant association between GRI pressing issues and the high profile of sustainability reporting, highly visible companies often have GRI performance. There is a preliminary study from the Australian Stock Exchange (ASX) Top 100 companies to investigate the state of sustainability report assurance.

There has been ongoing discussion of the methods most suitable for providing assurance of sustainability reports. In general, the assurance of sustainability reports is concerned with all the information within a report, the bounds of performance auditing. Thus assurance is considered the most appropriate means to provide a degree of confidence in the accuracy and completeness of the reported information (Australia and New Zealand & Balance Foundation, 2013). For example, the assurance of carbon footprint reports in the Australian context has utilised technical specialists to check the scope, corporate boundary, and calculation methodologies. In the case of the belated implementation of the rather limited carbon pollution reduction scheme, many voluntary schemes were initiated by industry to facilitate the measurement and auditing processes. Like any form of assurance, sustainability report assurance addresses the subject matter (the report) and attaches it to a certain level of assurance. The form and content of the assurance statement are the other components of sustainability report assurance. Therefore, the assurance statement can vary in scope, raising issues concerning comparable engagements.

8.2. Benefits and Challenges of Assurance

Review: Examines the advantages and obstacles of assurance.

Key Topics: Credibility, cost, and resource requirements.

Outcome: Provides a balanced view of assurance in GRI reporting.

External assurance of the quality of a corporate sustainability report is becoming a new customary in the corporate practice of reporting sustainability performance. The Sustainability Committee in GRI recently has been confronted with an increasing interest in assurance issues from reporters, readers, and assurance providers. Nevertheless, while external assurance is put forth as a mean to increase the quality of information, it still remains unclear in what way this transparency tool translates into an improvement of the overall quality of information disclosed. Looking at the reasons why the sustainability reports of the largest listed companies need to be assured underpin this analysis that external assurance of sustainability reports is mainly needed where the business drives substantial social and environmental impacts and risks to stakeholders and what is publicly disclosed is used by stakeholders as an essential part of informed decision-making or in their accountability relations with company executives. In this case, it could be expected that for sectors where this link between the business and the relevance of sustainability disclosure is particularly strong there should be much more demand for assurance and assurance statement should have a stronger positive impact on the quality of disclosed information. Having this broad frame in mind, this piece of work tries to examine whether the sectoral affiliations of companies drive either the adoption of assurance statements about the quality of disclosed information and the overall quality of the information contained in disclosed reports (Achmad et al., 2017). This paper conducts an analysis of the largest listed companies that are asked to disclose their information on the basis of the global reporting initiative guidelines, so as to look at a sample that meets common disclosure requirements both for the size and the nature of the market they operate in.

For many years an increasing concerns has been expressed about the trustworthiness of the vast array of information that is emerging regarding social and environmental management and describes the impacts of business activity (Faisal, 2017). Recently, the voluntary commitment has been made by a number of companies to external verification and independent assurance, in order to reinforce the reliability and credibility of social and environmental reports. As a result, a set of worked-based intensive tasks has been developed and is now widely offered by the multiple assurance providers. A literature has meaningfully advanced in this matter, showing the extent that these transparent assurance services are taken up by the companies and the impact on report readers.

8.4. Challenging Questions and Answers

Q1: What are the main types of assurance in GRI reporting, and how do they differ?

A1: The main types are internal and external assurance. Internal assurance is conducted by the organization itself, while external assurance involves independent third parties. External assurance is generally considered more credible due to its objectivity.

Q2: Why is external assurance becoming more common in corporate sustainability reporting?

A2: External assurance is increasingly adopted to enhance the credibility and reliability of sustainability reports, particularly for companies with significant social and environmental impacts. It helps stakeholders make informed decisions and holds companies accountable.

Q3: What challenges do organizations face when implementing assurance for GRI reports?

A3: Challenges include high costs, resource intensity, and the complexity of ensuring comprehensive and accurate reporting. Additionally, the lack of standardized assurance practices can lead to variability in the quality of assurance statements.

Q4: How does assurance impact the quality of sustainability reports?

A4: Assurance can improve the accuracy and completeness of reported information, but its direct impact on overall report quality remains debated. It depends on the rigor of the assurance process and the commitment of the reporting organization.

Q5: What role does sectoral affiliation play in the adoption of assurance statements?

A5: Companies in sectors with significant social and environmental impacts are more likely to adopt assurance statements, as stakeholders in these sectors demand higher transparency and accountability.

Q6: How does GRI reporting address governance gaps in sustainability management?

A6: GRI reporting serves as a transnational governance tool that complements public regulations, encouraging companies to adopt "beyond-compliance" behaviors and address gaps in formal legislation.

8.5. Summary

Chapter 8 explores the role of assurance in GRI sustainability reporting, emphasizing its importance in enhancing report credibility and reliability. It discusses various assurance methods, including internal and external approaches, and their benefits, such as improved transparency and stakeholder trust. However, challenges like cost, resource requirements, and variability in assurance quality are also highlighted. The chapter underscores the growing adoption of external assurance, particularly in sectors with significant social and environmental impacts, while questioning its direct impact on overall report quality. It also examines how GRI reporting addresses governance gaps by encouraging progressive corporate behavior beyond regulatory compliance.

Chapter 9

GRI and Integrated Reporting

Review: Explores the relationship between GRI and integrated reporting.

Key Topics: Holistic reporting, financial and non-financial integration.

Outcome: The chapter highlights the complementary relationship between GRI and integrated reporting, emphasizing how GRI's focus on sustainability reporting aligns with the holistic, long-term value creation approach of integrated reporting. It demonstrates that GRI's materiality-driven framework and integrated reporting's connectivity principles work synergistically to enhance transparency and accountability in corporate reporting.

9.1 Introduction

In South Africa, the Global Reporting Initiative (GRI) is commonly regarded as the bedrock of responsible and sustainable investment and reporting (Gibassier et al., 2016). South African companies, particularly large corporations of British origin, such as Old Mutual, Anglo-American, and Nedcor, turned to sustainability and social reporting as a tool to shed their tarnished image as protégés of the apartheid regime and enhance their reputation abroad. Although social accountability emerged during the period of the apartheid regime, the conditions for its flowering did not exist in the confrontation between two virtually antagonistic and mutually exclusive camps. Nonetheless, the GRI is generally commended for its theoretically underpinned vision and the positive impulse to integrate economic, environmental, and social sustainability. Although the number of published sustainability reports is still low compared with financial reports, the share of global non-financial or combined reports mainly produced in accordance with the GRI guidelines is steadily rising. A set of variables such as a strong commission for sustainable development, a sound guiding financial policy to internalize external costs of production and consumption, integrated sustainability and quality management, certification by international standards covering environmental protection and workplace conditions, and conditional managerial capitalism with active boards of directors overseeing executive pay and carrying out risk and crisis prevention can be seen to exert a manifold influence on each other.

9.2. Concept of Integrated Reporting

Review: Introduces the principles of integrated reporting.

Key Topics: Value creation, connectivity, and long-term thinking.

Outcome: Clarifies the purpose and benefits of integrated reporting.

Business has been actively promoting Sustainable Development Goals (SDGs) in the recent couple of years. In line with this movement, an essential tool in promoting sustainability will be the Global Reporting Initiative (GRI) issuances. It is the most used voluntary sustainability reporting standard, obtaining widespread adoption in all continents. Consequently, GRI is the best reflector of sustainability reporting practices carried out globally. However, uses of GRI data in global scale has been limited, and analyses relying on such data would provide valuable insights to both the businesses and the wider community.

Financial accountants have long been recognized as key players in corporate sustainability efforts. Their awareness of the impact of corporate activities on sustainable development grew in light of widespread corporate misconducts. Therefore, it was suggested that they could play a role in assuring the compliance of economic, social and environmental reporting with the Global Reporting Initiative (GRI) guidelines and assist in GRI report assurance. But in light of the recent scholarly critique raised against the GRI assurance model, further developments are needed in the field. Avenues for accountants' involvement in sustainability and the chances and barriers to such involvement are explored further in the future environmental, social and governance (ESG) assurance scheme modeled after the Integrated Reporting (IIRC) Framework (Gibassier et al., 2016).

9.3. Synergies between GRI and Integrated Reporting

Review: Examines how GRI and integrated reporting work together.

Key Topics: Shared principles, complementary frameworks.

Outcome: Demonstrates the synergies between the two reporting approaches. There is a natural synergy between GRI and Integrated Reporting as GRI was the trailblazer for Integrated Reporting by being the first global reporting quality regime on company's environmental and social impacts, and co-found SustainAbility in 1987. The focus of GRI on reporting externalities through a Lens of Materiality goes hand in hand with the scopes of issues, risks and opportunities that should be disclosed in Integrated Reporting's structure.

With G4 the main goal was to shift to a structure of reporting that is based on the remits to disclose certain issues, risks and opportunities companies face. G4 was also tied with the creation of the first International Integrated Reporting Council (IIRC) structure, and later, in 2013, a recalibration of G4's Guidelines and new structure features were born, such as the Disclosures on Management Approach (DMAs). This means companies that report through GRI have the same C-suite binded mandate to comply with IR principles. By 2013 the IIRC had certified the G4 Frameworks as adhering to IR principles. With this there was a Vicious Circle GRI-IIRC push to companies that report on their environmental and social transparency to do IR and vice versa (Goldman, 2015).

9.4. Challenging Questions and Answers

Q1: How does GRI support the principles of integrated reporting?

A1: GRI provides a robust framework for reporting on environmental, social, and governance (ESG) issues, which aligns with integrated reporting's focus on holistic value creation and long-term thinking. GRI's materiality approach ensures that key sustainability issues are integrated into corporate reporting.

Q2: What are the key synergies between GRI and integrated reporting?

A2: Both frameworks emphasize transparency, accountability, and the integration of financial and non-financial information. GRI's focus on materiality and externalities complements integrated reporting's emphasis on connectivity and value creation over time.

Q3: Why is GRI considered a trailblazer for integrated reporting?

A3: GRI was one of the first global frameworks to standardize sustainability reporting, setting the stage for integrated reporting by emphasizing the importance of disclosing environmental and social impacts alongside financial performance.

Q4: How does integrated reporting enhance the value of GRI reports?

A4: Integrated reporting provides a broader context for GRI data by connecting sustainability performance to overall business strategy and long-term value creation, making GRI reports more relevant to stakeholders.

Q5: What challenges arise when integrating GRI and integrated reporting?

A5: Challenges include aligning different reporting frameworks, ensuring consistency in data, and managing the complexity of combining financial and non-financial information in a cohesive manner.

Q6: How has GRI evolved to align with integrated reporting principles?

A6: GRI's G4 guidelines introduced features like Disclosures on Management Approach (DMAs), which align with integrated reporting's focus on connectivity and materiality, fostering greater synergy between the two frameworks.

9.5. Summary

Chapter 9 explores the relationship between GRI and integrated reporting, highlighting their complementary nature. GRI's focus on sustainability reporting aligns with integrated reporting's holistic approach to value creation, emphasizing transparency, accountability, and long-term thinking. The chapter discusses how GRI's materiality-driven framework and integrated reporting's connectivity principles work together to enhance corporate reporting. It also examines the evolution of GRI, particularly through its G4 guidelines, to align with integrated reporting principles, fostering greater synergy between the two frameworks. The chapter underscores the importance of integrating financial and non-financial information to provide a comprehensive view of corporate performance.

Chapter 10
GRI and Industry-Specific Reporting

Review: Discusses the adaptation of GRI for specific industries.

Key Topics: Sector supplements, tailored indicators.

Outcome: The chapter emphasizes the adaptability of GRI frameworks to industry-specific needs through sector supplements and tailored indicators. It highlights how GRI addresses unique challenges in various industries, such as mining, oil and gas, and banking, by involving multi-stakeholder groups to identify material sustainability issues. This customization enhances the relevance and effectiveness of sustainability reporting across diverse sectors.

10.1 Introduction

As a global network based on the vision of a sustainability economy, the GRI contributed firstly by developing a framework that organisations could use to report their sustainability progress. A GRI report provides a common structure in which economic, environmental, and social impacts are reported. Whilst reporting on issues particular to each industry, the reports have commonalities and industries do benefit from specific vs. generic reports. The GRI has supported this industry-specific reporting. Jointly with the United Nations Environment Programme, there were pilots in six areas: banking, electricity utility, food processing, mining, oil and gas, and paper products. The GRI has now released its Sector Supplements to four of these industry sectors. It brought together large stakeholders in the sectors to identify the most material sustainability indicators, thereby ensuring a more effective process. In so doing, these sectors now have customised reporting guidelines. The October 2002 International Mining and Metals Conference held in Johannesburg was devoted to sustainably development and, amongst others, looked at the role of the GRI Mining Supplement. This was the second conference dedicated to industry-specific guidelines and it has now become annual practice to have such conference. While the Mining Supplement has been published with community consultation, the others will be available for public comment. Commercial mining companies, labour unions, government, and community groups make up each of the multi-stakeholder technical advisory groups. Former South African President addressed the October 2002 Conference. Australia and Canada will be using the GRI Mining Supplement and, with such large mining industries in both countries, it is a significant achievement. Unchecked consumption is costing the planet much more than any savings and, indeed, contributing to the high levels of inequality. Rising from the shipwreck of the United Nations Conference on Sustainable Development, there have been global queries as to whether global civil society is able to overcome this. Some very powerful people are not committed to sustainability. On the 3rd day of the EU industrial leaders in Brussels, the Sustainability Game was energised. Millions of people cannot be brought out in a million protests. Social Audit may not yet have teeth but does have a megaphone. Beyond the stakeholders, civil society discussion changes the mood. The lessons from the campaign were: don't argue with data, make effective propaganda moves, and engage the key players directly.

10.2. Sector Supplements and Specific Standards

Review: Explores GRI's industry-specific guidelines.

Key Topics: Customization, relevance, and applicability.

Outcome: Shows how GRI addresses unique industry challenges.

The Sustainability Reporting Guidelines, in the form of sector supplements, or other specific standards, are currently available only for certain industry sectors, while for the rest, there are no specific standards. Thus, coverage of some sectors of economic activities is more extensive due to the fact this kind of reporting has long-standing tradition, initiative of companies within certain industry sectors, or pressure by large financial institutions that the corporations financed through loans or equity investment in the sectors that are, by their understanding, riskier for the environment and human health. These standards have been developed in multi-stakeholder approach, meaning that international, independent Working Groups are developing them. As any other kind of standard, they are a systematic set of criteria used as reference of measurement. Additionally, standards can be more or less obligatory for the companies active in particular industry sector; more reputable and prestigious, the ones developed by internationally recognized institutions, therefore more in confidence in their application, attention, acceptance by the customers and buyers, but also protection of the producers and consumers.

10.3. Examples of Industry-Specific Reporting

Review: Provides examples of GRI reporting in various sectors.

Key Topics: Case studies, best practices.

Outcome: Illustrates successful applications of GRI in different industries.

Among those corporations voluntarily reporting according to the Global Reporting Initiative's (GRI) Sustainability Reporting Guidelines in a selection of Emerging Asian economies, the coverage of the guidelines is addressed, in addition to examination of the sectors in which reporting is most prevalent. There is a significant increase in the rate of sustainability reporting that involves the GRI guidelines, suggesting that they have made a difference in promoting this practice (Romana, 2018). Comparisons are made between transparency of different industry sectors and trend of disclosures over time in two economies, China and Thailand. Lifting the overall standards of responsible business practice, and hence of sustainable development in this region, clearly depends in part on a better understanding of, and responsiveness to, business behavior. A step in this direction would be policies to promote and monitor the effectiveness of business sustainability reporting. A survey of sustainability reports by 150 companies in ten Emerging Asian economies showed that (the geographical breakdown by number of companies in the survey is): China, Hong Kong and Taiwan (50 companies); South Korea (20); Thailand (10); Indonesia and Philippines (5); and India, Malaysia and Singapore (number of companies reporting). Voluntary corporate disclosure of environmental and social responsibility activities has been growing in this region in line with trends elsewhere in the world, as has interest in the use of sustainability reporting by corporations (Wu et al., 2017). Seven information facilities were used to identify companies potentially reporting in the economies surveyed....Being comparable, the large data sets for China, Hong Kong and South Korea will be used to compare the sectors in which companies are reporting. Additionally, focusing on China and Thailand, a comparison is made between the transparency of reporting by companies in the same sectors. Changes in the amount of reporting over time also are examined in Thailand. The latter comparison is of particular interest given that much analysis of the determinants of sustainability reporting within particular countries suggests that competition can be a driving force in its take up.

10.4. Challenging Questions and Answers

Q1: What are GRI Sector Supplements, and why are they important?

A1: GRI Sector Supplements are industry-specific guidelines that tailor sustainability reporting to address unique challenges and material issues in sectors like mining, oil and gas, and banking. They ensure relevance and effectiveness in reporting.

Q2: How does GRI ensure the relevance of its industry-specific guidelines?

A2: GRI collaborates with multi-stakeholder groups, including industry experts, governments, and NGOs, to identify material sustainability indicators and develop customized reporting frameworks for each sector.

Q3: What industries have benefited most from GRI's sector-specific reporting?

A3: Industries such as mining, oil and gas, banking, and food processing have benefited significantly, as these sectors face substantial environmental and social risks that require tailored reporting.

Q4: How does GRI's multi-stakeholder approach enhance its credibility?

A4: By involving diverse stakeholders in the development of sector supplements, GRI ensures that its guidelines are comprehensive, balanced, and reflective of the concerns of all relevant parties.

Q5: What challenges arise in implementing industry-specific GRI reporting?

A5: Challenges include aligning diverse stakeholder interests, ensuring consistency across industries, and adapting guidelines to evolving industry practices and regulations.

Q6: How has GRI influenced sustainability reporting in emerging economies?

A6: GRI has driven increased sustainability reporting in emerging economies like China and Thailand, promoting transparency and responsible business practices through its widely adopted guidelines.

10.5. Summary

Chapter 10 explores GRI's adaptation to industry-specific reporting through sector supplements and tailored indicators. It highlights the importance of customizing sustainability reporting to address unique challenges in sectors like mining, oil and gas, and banking. The chapter discusses GRI's multi-stakeholder approach, which ensures the relevance and credibility of its guidelines by involving industry experts, governments, and NGOs. Examples from emerging economies, such as China and Thailand, illustrate the growing adoption of GRI reporting and its impact on promoting transparency and responsible business practices. The chapter underscores the flexibility of GRI frameworks in addressing diverse industry needs while fostering global sustainability reporting standards.

Chapter 11

GRI and Small and Medium Enterprises (SMEs)

Review: Examines how SMEs can adopt GRI reporting.

Key Topics: Adaptation, resource constraints, and benefits.

Outcome: The chapter emphasizes the importance of GRI reporting for SMEs, highlighting its potential to enhance transparency and sustainability practices. While SMEs face challenges like resource constraints and complexity in adopting GRI, tailored frameworks and simplified guidelines can make reporting more accessible. The chapter underscores the collective environmental and social impact of SMEs and the need for their inclusion in global sustainability efforts.

11.1 Introduction

The Global Reporting Initiative now envisions a sustainable economy and world where sustainable decision making and reporting is standard practice. However, such an ambition requires rapid advances in the approach to sustainability, particularly in regard to radical innovation, ecosystem protection, and the use of systems thinking in strategic planning regarding interconnected stakeholder and system considerations. In this review, the innovative aspects of the fourth generation sustainability report are noted and evaluated with respect to those developments needs. While this edition retains a sustainability perspective, it goes beyond most previous reports by operationalising the Eco-Indicator lifecycle assessment method in order to enable a complete ecological sustainability analysis of an apartment building. The resulting calculations informed innovative final report design, including a number of data visualisations like the apartment's resource flows. These flow diagrams also inform the report's systems perspectives by illustrating how resources are used along and across life cycles, spatial scales, and sectors. Meanwhile, an innovative modelling step—an apartments in/apartments out matrix—enables the assessment of a dual-unit's relative share in the resources used and pollutants released by the entire building.

11.2. Adaptation of GRI for SMEs

Review: Explores how GRI can be tailored for SMEs.

Key Topics: Simplified frameworks, scalability.

Outcome: Provides guidance for SMEs to implement GRI reporting effectively.

Sustainability reporting continues to be a focus in ensuring organizational transparency and accountability to internal and external stakeholders. The growth in the number of companies issuing sustainability reports in recent years has primarily been catalyzed by the appearance of various voluntary sustainability reporting guidelines and principles. As a leading example, the Global Reporting Initiative (GRI) framework has been used by over 4,000 reporting organizations globally (van der Walt, 2018).

Small-to-medium sized enterprises (SMEs) can exert considerable pressure on the environment and society just like large organizations. While the pressures and impacts derived from their actions are smaller in scale, SMEs are collectively numerous and, therefore, their overall impact can be large. There is some evidence that the pressures from SMEs can be substantial because of business practices chosen to reduce costs, their vulnerability to economic and regulatory changes, and increasing consumer pressure. The motives for investigating SMEs stem from the consideration that the health and wellbeing of any community is dependent on the health and sustainability of a full network of enterprises, from the village store to large manufacturing plants, and that only comprehensive knowledge of sustainability practices within this network would enable an integrated comparative analysis across scales and sectors.

There is evidence that some SMEs do report on their sustainability impacts using the GRI framework; however, others have found that it is hardly applicable for SMEs. By using the GRI framework, organizations are required to report on a large number of sustainability indicators and this can render report production expensive and very time consuming for SMEs. Moreover, comparably concise sustainability indicators are missing in the GRI framework. Overall, it appears that even when small and simple guidelines exist, SMEs face specific barriers that limit their capacity to prioritize key issues and take advantage of environmental strategies.

11.3. Benefits and Challenges for SMEs

Review: Discusses the advantages and obstacles for SMEs using GRI.

Key Topics: Cost, capacity, and stakeholder engagement.

Outcome: Offers a balanced perspective on GRI adoption for SMEs.

The adoption of sustainability strategies by Small and Medium-sized Enterprises (SMEs) is critically discussed in terms of financial results, benefits, risks, and barriers. The European Commission defines SMEs as those enterprises which employ fewer than 250 people, and have an annual turnover not exceeding 50 million euro, and/or an annual balance sheet total not exceeding 43 million euro. Globally, they are the most numerous type of enterprises, share the vast majority of employment, comprising around half of the Gross Value Added, involve a variety of sectors, and are seen as a source of employment. Due to the heterogeneity of their nature, the support and engagement of SMEs within a broader sustainability strategy can be particularly important. Frequently perceived as "local players," SMEs are often deeply rooted in their territory. Stimulating their engagement in projects and initiatives with a common sustainable vision can indeed foster territorial development, promoting environmental and social progress in the area. Additionally, as part of the supply chain, SMEs can exert considerable pressure on the environment and society, placing demands on compliance with standards by upstream partners. On the other hand, this group of organizations is often subject to a network of strong pressures, increasing their vulnerability. The financial burden of compliance with rules, the increasing expectations of the stakeholders, together with the fierce competition are but a few of the factors which constitute a considerable challenge for these organizations. Hence, it is possible to assert that SMEs' sustainability strategies should become a sound interest of both policy makers and stakeholders (van der Walt, 2018).

11.4. Challenging Questions and Answers

Q1: Why is GRI reporting relevant for SMEs despite their smaller scale?

A1: SMEs collectively have significant environmental and social impacts due to their large numbers. GRI reporting helps them improve transparency, accountability, and sustainability practices, contributing to broader global sustainability goals.

Q2: What are the main barriers SMEs face in adopting GRI reporting?

A2: SMEs often struggle with resource constraints, high costs, and the complexity of GRI's extensive reporting requirements, making it challenging to prioritize sustainability reporting.

Q3: How can GRI frameworks be adapted for SMEs?

A3: Simplified and scalable GRI frameworks, tailored to SMEs' capacities, can make reporting more accessible. This includes reducing the number of indicators and providing clear, concise guidelines.

Q4: What benefits do SMEs gain from adopting GRI reporting?

A4: Benefits include improved stakeholder trust, enhanced reputation, better risk management, and alignment with global sustainability standards, which can lead to competitive advantages.

Q5: How do SMEs contribute to broader sustainability goals?

A5: SMEs, as key players in local economies and supply chains, can drive environmental and social progress by adopting sustainable practices, influencing larger networks, and fostering community development.

Q6: What role do policymakers play in supporting SMEs' sustainability efforts?

A6: Policymakers can provide incentives, resources, and simplified frameworks to help SMEs overcome barriers and integrate sustainability into their operations effectively.

11.5. Summary

Chapter 11 explores the relevance of GRI reporting for SMEs, emphasizing their collective environmental and social impact. While SMEs face challenges like resource constraints and complexity in adopting GRI, tailored and simplified frameworks can make reporting more accessible. The chapter highlights the benefits of GRI adoption, such as improved transparency, stakeholder trust, and alignment with global sustainability goals. It also underscores the importance of SMEs in driving local and global sustainability efforts and calls for policy support to help SMEs overcome barriers and integrate sustainable practices

Chapter 12

GRI and Climate Change Reporting

Review: Focuses on GRI's role in climate change reporting.

Key Topics: Emissions, risks, and opportunities.

Outcome: The chapter underscores GRI's critical role in enabling organizations to report on climate change impacts, risks, and opportunities. It provides guidance on measuring and disclosing greenhouse gas (GHG) emissions, aligning with global frameworks, and improving transparency. The chapter highlights the growing trend of businesses adopting GRI standards to enhance their sustainability reporting and address climate challenges effectively.

12.1 Introduction

In addition to transparency in economic, social and management aspects, it is important to make public the financial impact related to climate change; the associated risks and opportunities; the information and relationships related to initiatives undertaken by the entity in this regard, as well as the adjustment of financial plans related to risk. Furthermore, it is stated that the greenhouse gas (GHG) emissions expressed on behalf of governments such as CO2e as a separate amount are requested. Therefore, in the guidelines, it is necessary to publicly disclose the data related to GHG emissions (tCO2e) in order to make a distinct and separate statement on behalf of countries. No reference is available to the public regarding the GHG emissions (tCO2e) of the entity in the document or in the digital environment. In addition to the national emissions, data associated with market-based and location-based emissions are not available to the public. Sustainability leaders are increasingly required to be transparent in terms of internal issues, uncertainties and credibility concerns. Therefore, the entity should make a statement about the level of confidence of the data disclosed and the calculation methods used. And also, there is no explanation regarding what the GHG emissions are related to the company's activities.

The results indicate that all the companies examined make disclosures related to the guidelines. However, the levels of disclosure remain incomplete. Furthermore, the result indicates that the majority of the disclosures are made regarding the water-related topics. Finally, it is suggested that a standard format for the disclosure of sustainability reports is not prescribed by the guidelines; this might be the reason why the level of disclosure is low.

12.2. Guidance on Climate Change Reporting

Review: Provides guidance for reporting on climate issues.

Key Topics: Metrics, disclosure, and alignment with global frameworks.

Outcome: Helps organizations effectively report on climate-related impacts.

There is a strong trend among global businesses in the era of the Anthropocene to portray themselves as sustainable in a material way. Each year, more than 90% of the 250 largest global businesses produce reports intended to depict them as sustainable. In the period from 2005 to 2017, the number of the world's largest businesses which have provided online sustainability reports has increased eightfold from 35 to 285. In most sustainability reports, greenhouse gas (GHG) emissions are considered to be particularly important for the business sector and are reported, with two thirds of reports mentioning them. There are more than fifty instruments issued by international organizations that determine how the GHG emissions of a company should be reported (Smit & van Zyl, 2016). There's a rapidly increasing tendency for companies to attempt to comply with those standards. In 2018, 48% of the 635 companies which have disclosed their GHG emissions have responded to one of those inquiries. Major transnational corporations build their image on sustainability and adopt the "Global Reporting Initiative" reporting standard, which is considered to be the best one. There are numerous studies which analyze either the quality of the information provided in reports or GHG emissions by business in various countries under an influence of standards on reporting, since the issuing of the Kyoto Protocol. There have only been isolated studies that analyze either the quality of the information provided in reports or GHG emissions by the businesses of specific states and that have recently appeared.

12.3. Carbon Footprint Measurement and Reporting

Review: Explores methods for measuring and reporting carbon footprints.

Key Topics: Calculation, transparency, and reduction strategies.

Outcome: Highlights the importance of accurate carbon reporting in GRI.

The diverse – and global– stakeholders of organisations require their widely used goods and services to be created in a sustainable manner. As citizens grow more aware of the scarcity and endlessness of resources of the planet, businesses are expected to improve their practices accordingly. Therefore, companies strive to publicize their social and environmental effectiveness and pledge to fulfil their duties or even outperform legislative requirements wherever possible.

One of the most important, internationally active initiatives in this field is the Global Reporting Initiative (GRI) launched in 1997. Established by a collaboration between the Coalition for Environmentally Responsible Economies (CERES) – a coalition of 150 groups of investors and authorities– and the United Nations Environment Program the initiative produces a sustainability reporting guide. Such standards help organisations understand and communicate more easily their environmental, social and commercial sustainability activities. GRI is further recognized by the OECD and the Holding Companies and Investment Business Ltd Guidelines as a general system for comprehensive sustainability reporting. The GRI enables consumers and other organisations to equitably consider the creation and assessment of sustainability reports and follow up with emerging developments and improvements. There is proof that the GRI's implementation is associated with improved sustainability reporting quality. The researchers found a rise in compliance with the GRI requirements following FY2003. There is further proof that the beginnings of the financial statements, particularly the KPIs, show a substantial improvement in disclosure quality. The largest improvement was seen in 2004 annual reports, although a long-term increase was also noted (Smit & van Zyl, 2016).

13.4. Challenging Questions and Answers

Q1: Why is climate change reporting important for organizations?

A1: Climate change reporting helps organizations disclose their environmental impacts, manage risks, and identify opportunities for sustainable practices, enhancing transparency and accountability.

Q2: What role does GRI play in climate change reporting?

A2: GRI provides standardized guidelines for measuring and reporting GHG emissions, enabling organizations to align with global frameworks and improve the quality of their sustainability disclosures.

Q3: What challenges do organizations face in reporting GHG emissions?

A3: Challenges include ensuring data accuracy, selecting appropriate calculation methods, and maintaining transparency in disclosures, especially for market-based and location-based emissions.

Q4: How does GRI improve the quality of sustainability reports?

A4: GRI's standardized framework ensures comprehensive and consistent reporting, leading to improved disclosure quality and better alignment with global sustainability goals.

Q5: What are the benefits of aligning climate change reporting with global frameworks?

A5: Alignment enhances credibility, facilitates comparability across organizations, and ensures compliance with international standards like the Kyoto Protocol and Paris Agreement.

Q6: How can organizations reduce their carbon footprint effectively?

A6: Organizations can adopt accurate measurement methods, set reduction targets, implement sustainable practices, and transparently report progress using frameworks like GRI.

13.5. Summary

Chapter 12 highlights GRI's role in climate change reporting, emphasizing the importance of transparency in disclosing GHG emissions, risks, and opportunities. It provides guidance on aligning reporting with global frameworks and improving the accuracy and quality of sustainability disclosures. The chapter notes the growing trend of businesses adopting GRI standards to enhance their sustainability practices and meet stakeholder expectations. It also explores methods for measuring and reducing carbon footprints, underscoring the need for accurate and transparent reporting. By leveraging GRI's framework, organizations can effectively address climate challenges and contribute to global sustainability efforts.

Chapter 13
GRI and Human Rights Reporting

Review: Examines the integration of human rights in GRI reporting.

Key Topics: Due diligence, accountability, and impact assessment.

Outcome: The chapter highlights GRI's role in promoting transparency and accountability in human rights reporting. It provides a framework for integrating human rights due diligence, impact assessments, and stakeholder engagement into sustainability reports. Case studies demonstrate how GRI guidelines enhance organizational transparency and trust, particularly in labor practices and human rights issues.

13.1. Introduction

GRI reporting 513 companies, with 486 companies grouped into 43 industries by a classifier, the Global Industry Classification Standard. Of these, 356 companies excluded 'none' core businesses, 'c' class companies with low market capitalization values, 'defunct' and 'bankrupt' companies, and data are extracted for rest 130 companies. An event study identifies abnormal returns of 130 companies around the time of releasing their GRI reports. Thus, this paper complements existing studies by documenting both improved performance and valuation of strategic corporate social responsibility initiatives materialized through GRI reporting, as well as by investigating such phenomena with a much larger and more recent sample compared to the prior literature.

13.2. Incorporating Human Rights in GRI Reporting

Review: Explores how human rights are addressed in GRI Standards.

Key Topics: Indicators, disclosure, and stakeholder engagement.

Outcome: Provides a framework for reporting on human rights issues.

By 2013, user companies in home countries of Exemplar companies incorporated Human Rights into reporting in line with Global Reporting Initiative (GRI) guidelines. By 2013 it will be apparent at a glance how companies responded to UN Principles on Business and Human Rights. One reason why Exemplar Companies' home governments were chosen was because to date most reputation damage to companies as a result of adverse UN procedures happened to them in home jurisdictions. There are therefore likely to be civil society and legal demands on them around Companies/ Industry Issues. In addition, a comparison could then be made of compliance of Exemplar companies with GRI guidelines on Human Rights' related issues of Transparency, Civil Society, and Remedies/Openness – setup, Implementation and Auditing, and whether there is any relationship between that and home state governments' future implementation of UN principles (Parsa et al., 2018). Spill-over may be in the form of policy guidance suggested by initial moves towards compliance with the UN principles. On this latter point if, as a result of this research, a demand were to be identified for further research that must be sequenced to happen after 2013 then scholarly theory articles on the normative benefits of such sequencing would be a useful adjunct.

13.3. Case Studies on Human Rights Reporting

Review: Presents examples of human rights reporting in practice.

Key Topics: Best practices, challenges, and outcomes.

Outcome: Illustrates successful human rights reporting using GRI.

This section examines the labour practices perspective given it refers specifically to transparency, accountability, and adoption of corporate reporting principles by stakeholder groups, more so than the other two sections. Therefore, this critical perspective is important as it highlights the potential failure as well as successes of GRI's role in promoting sustainability. It then covers possible legal action which attests to the trust in GRI is having a lasting and significant impact.

This reflects on the role of the Global Reporting Initiative (GRI) in promoting sustainability by examining its impact on human rights reporting. The analysis is informed by labour practices, health and safety, freedom of association, collective bargaining, anti-discrimination, and child labour. The case studies reveal a positive impact of applying the GRI guidelines in these settings. Following a short critical perspective on these findings, possible legal action in the EU is considered. This analysis shows that the GRI framework is having a significant and lasting impact on organizational transparency regarding human rights issues, thereby enhancing stakeholder trust (Parsa et al., 2018). The Global Reporting Initiative (hereafter GRI) is a networkbased organization which seeks to support sustainable development by promoting transparency and corporate accountability. The GRI launched its sustainability reporting guidelines in 2000 and they have since been adopted and used by many of the world's largest companies, non-business organizations, and other agencies.

13.4. Challenging Questions and Answers

Q1: How does GRI support human rights reporting?

A1: GRI provides specific indicators and guidelines for reporting on human rights issues, such as labor practices, anti-discrimination, and child labor, ensuring transparency and accountability.

Q2: What are the key challenges in integrating human rights into GRI reporting?

A2: Challenges include ensuring accurate data collection, addressing complex supply chain issues, and meeting stakeholder expectations for transparency and accountability.

Q3: How do GRI guidelines align with the UN Principles on Business and Human Rights?

A3: GRI guidelines align with the UN Principles by emphasizing due diligence, transparency, and remedies for human rights violations, helping companies comply with international standards.

Q4: What are the benefits of human rights reporting for organizations?

A4: Benefits include enhanced stakeholder trust, improved reputation, better risk management, and alignment with global sustainability and human rights standards.

Q5: How do case studies demonstrate the impact of GRI on human rights reporting?

A5: Case studies show that GRI guidelines improve transparency in labor practices, health and safety, and anti-discrimination, leading to greater accountability and stakeholder confidence.

Q6: What role does stakeholder engagement play in human rights reporting?

A6: Stakeholder engagement ensures that human rights issues are identified, addressed, and reported accurately, fostering trust and collaboration between organizations and their stakeholders.

13.5. Summary

Chapter 13 explores the integration of human rights into GRI reporting, emphasizing transparency, accountability, and due diligence. It provides a framework for reporting on human rights issues, such as labor practices, anti-discrimination, and child labor, aligning with the UN Principles on Business and Human Rights. Case studies illustrate the positive impact of GRI guidelines on organizational transparency and stakeholder trust. The chapter highlights the challenges of accurate data collection and supply chain complexities while underscoring the benefits of human rights reporting, including improved reputation and risk management. By leveraging GRI's framework, organizations can enhance their sustainability practices and contribute to global human rights efforts.

Chapter 14
GRI and Gender Equality Reporting

Review: Focuses on gender equality indicators in GRI Standards.

Key Topics: Diversity, inclusion, and empowerment.

Outcome: The chapter emphasizes GRI's role in advancing gender equality through standardized reporting. It highlights specific GRI indicators for gender equality, such as pay equity, representation, and workplace policies, enabling organizations to track and disclose progress. By promoting transparency and accountability, GRI reporting drives gender equality initiatives and aligns with global sustainability goals.

14.1. Introduction

Sustainability has become a watchword in the business world, and business houses can no longer exist without a strategy designed to mitigate their environmental impact. Over the last few decades, the concerns of sustainable development and corporate social responsibility have metamorphosed into the context of business strategy, which is having significant implications for society and the economy. The corporate sector is now, therefore, required to discharge new responsibilities towards consumers, employees, investors, and the wider community. The corporations, on the other hand, are rushing to incorporate sustainability in their business models and adopt the triple-bottom-line approach. Compliance with reporting guidelines, either statutory or voluntary, has become mandatory to communicate to the stakeholders how corporations, in reality, treat the eco-system. Equity, fairness, justice are the spine words of corporate governance and because of this, companies in their pursuit of long-term survival and growth are nowadays integrating humanitarian outlook into their strategies. (Ghoogassian, 2015). On the other hand, corporations, as far as practicable, should also come out with gender-equality related informative declarations.

14.2. Gender Equality Indicators in GRI Standards

Review: Explores specific GRI indicators for gender equality.

Key Topics: Pay equity, representation, and workplace policies.

Outcome: Provides tools for organizations to report on gender equality.

All new General Standard Disclosures are written in normal font, while all new Specific Standard Disclosures are written in italicized font. General Standard Disclosures apply to all organizations, independent of their identification of Material Aspects.

Entity 102-1 refers to the name of the organization, 102-2 covers its activities, brands, products, and/or services, and 102-3 specifies the location of its headquarters. Questions such as the CEO's gender, the number of women in the C-suite, and plans to address the gender pay gap are critical. Short-term strategic decisions can significantly influence long-term value creation and relationships with stakeholders, including investors, consumers, suppliers, employees, policymakers, and business partners (Wuerzinger, 2015). In 1999, two years of consultations across five countries were conducted to prepare the first draft of the GRI guidelines, which were released for public feedback. The initial version was launched in New York in 2000, with subsequent updates in 2002, 2006, and 2011. The 2006 and 2011 versions were mandatory for organizations classified as "GRI Associate" users.

Entity 102-6 outlines the organization's structure, including its main divisions, operating companies, subsidiaries, and joint ventures, while 102-9 addresses the impacts of its activities, products, and services on biodiversity. GRI Standard Disclosures are applicable once an organization identifies a Material Aspect, Material Issue, or Key Matter. Defining these material topics is a rigorous process, as they are intended to reflect the organization's significant economic, environmental, and social impacts or those that substantially shape stakeholder assessments and decisions (GRI Research and Development Atelier of Sustainability Reporting Practice, 2012:7.9.a), as outlined in the "G4 Guidelines."

14.3. Promoting Gender Equality through Reporting

Review: Discusses how reporting can drive gender equality initiatives.

Key Topics: Transparency, accountability, and progress tracking.

Outcome: Shows the impact of reporting on promoting gender equality.

In the context of the Sustainable Development Goals, the gender equality sustainability aspect should be analyzed. Specifically, the gender equality aspect of non-financial reporting should now be conceived by analyzing the following research question: To what degree do companies report on gender equality aspects, and do gender equality-related indicators represent a useful extension of prior research frameworks to classify data from non-financial reports? In order to answer this question, an analysis of the gender equality aspect is presented within a large sample of sustainability reports collected according to established guidelines. This analysis provides avenues for further research into how gender equality can be more inclusively reported and thus permits a more comprehensive understanding of gender implications.

Globally, the established guidelines have become the most influential standard-setter and reporting organization for non-financial reports. The importance of a gender equality aspect of corporate non-financial reporting is also recognized and included in global surveys. Nonetheless, focus has not been placed on the gender equality pillar because it is still a relatively uncommon topic. However, at latest after major world events and their objective to "achieve gender equality and empower all women and girls," each individual company should be alive to the topic of gender equality in aspects of non-financial reporting. Non-financial reporting in general can be subdivided into environmental and social reporting; the latter comprises many different topics among which the gender equality seems to be low ranked within international guidelines. On the other hand, the problem of data with gender components may have been invisibly present because the general tendency in non-financial reporting is to overlook the separation of gender aspects.

14.4. Challenging Questions and Answers

Q1: What are the key gender equality indicators in GRI Standards?

A1: Key indicators include gender pay equity, representation in leadership (e.g., C-suite), workplace policies, and diversity metrics, helping organizations track and report on gender equality.

Q2: How does GRI reporting promote gender equality?

A2: GRI reporting promotes gender equality by requiring transparency in gender-related metrics, encouraging organizations to address disparities and implement inclusive policies.

Q3: Why is gender equality reporting important for organizations?

A3: Gender equality reporting enhances accountability, improves stakeholder trust, and aligns with global sustainability goals like SDG 5, fostering inclusive and equitable workplaces.

Q4: What challenges do organizations face in gender equality reporting?

A4: Challenges include data collection, ensuring accurate representation, and addressing systemic barriers to gender equality within organizational structures.

Q5: How does GRI align with the Sustainable Development Goals (SDGs)?

A5: GRI aligns with SDG 5 (Gender Equality) by providing frameworks for reporting on gender-related metrics, helping organizations contribute to global gender equality efforts.

Q6: What impact does gender equality reporting have on corporate culture?

A6: Gender equality reporting fosters a culture of inclusivity, transparency, and accountability, encouraging organizations to prioritize diversity and equity in their operations.

14.5. Summary

Chapter 14 focuses on GRI's role in advancing gender equality through standardized reporting. It explores specific GRI indicators, such as pay equity, representation, and workplace policies, enabling organizations to track and disclose progress on gender equality. The chapter highlights the importance of transparency and accountability in driving gender equality initiatives, aligning with global sustainability goals like SDG 5. It also discusses challenges in data collection and systemic barriers while emphasizing the positive impact of gender equality reporting on corporate culture and stakeholder trust. By leveraging GRI's framework, organizations can contribute to creating inclusive and equitable workplaces.

Chapter 15
GRI and Technology Innovation

Review: Examines the role of technology in GRI reporting.

Key Topics: Data collection, automation, and digital tools.

Outcome: The chapter highlights how technology enhances GRI reporting by improving data collection, automation, and real-time reporting. It demonstrates the benefits of digital tools, such as software and analytics, in streamlining reporting processes and increasing accuracy. Emerging trends like AI and blockchain are also explored, showing their potential to revolutionize sustainability reporting.

15.1. Introduction

From education to supervision and evaluation, the Global Reporting Initiative (GRI) plays an essential role in promoting sustainability in organizations internationally, is playing many roles on the sustainability platform, and is going to play an expanding and deepening substantial role in the process of sustainable development for the world. The Chinese case suggests that government intervention is important in the early establishment of the reporting framework to increase the numbers of reporters, but reporting should have an internal a long-term benefit, it points to the limited role of Chinese governments in pushing the quality and transparency of reports (Barkemeyer et al., 2015). The Government of China and other governments still have much room to improve their guidance on what information reporters should disclose. For technology innovation, GRI could join international cooperation and knowledge transfer concerning technology to help facilitate transfer of advanced LCT/KET to developing countries. It can also engage in technology capacity building, programmes and projects, and guideline development for sustainability technology reporting through elaborating transparent technological innovation and management strategies for technology-driven companies. Another avenue is working with stock exchanges to foster corporate sustainable report requirement and promote exchanges through monitoring the equality and management of information flow to push promoted and transparent traded price for sustainable technology-related product and services. Beyond technology innovation, GRI can work with policy markers to review policy inconsistencies that inhibit greater sustainable innovation. Ultimately, public policy and corporate supplier reporting practices need to shift from a focus solely on emission reduction and efficiency strategies to those policies and best practices that facilitate innovation in both products and processes at the firm level. From an international development cooperation revolving around climate change taken as a top-down challenge. GRI can promote response to and understanding of other bioproblems and adjusted risk and vulnerabilities to the developing country context in IT and or other sectors. Measures could include the building capacity on technology analysis other than energy and energy-efficiency and its use other than the one adopted by clean technology mechanism and the alignment among an analytical content of National Adaptation Plan, technology needs-assessment, and other comparative studies on technology choices.

15.2. Use of Technology in GRI Reporting

Review: Explores technological tools for GRI reporting.

Key Topics: Software, analytics, and real-time reporting.

Outcome: Demonstrates the benefits of technology in streamlining reporting processes.

Recently, companies have widely started to prepare sustainability reports and follow the rules of the Global Reporting Initiative (GRI) in these reports. According to a guideline formed by the GRI, companies report entirely on economic, social and environmental sides and provide consistency of these reports. It aims to collect public based information about companies, increase the competition between companies in the context of accountability, scan, analyse the companies in terms of sustainable development, give the signal of these works to the investors. These reports are written in the context of the GRI format and are published on the internet also. The use of the internet in the publication of these reports provides contributions towards convenient access to the reports. GRI reports located on the internet of some companies, since a tool using technology provides more factors to be investigated. Therefore developed for this survey procedure is tested on one part of the BIST-30 index, based on the sustainability reports published in 2012, and evaluations are made on the results. It is understood from the evaluation results that the procedure developed for this report has been successful and provides convenience in the investigation of the reports (Wuerzinger, 2015).

15.3. Emerging Trends in Technology and Reporting

Review: Discusses future trends in reporting technology.

Key Topics: AI, blockchain, and predictive analytics.

Outcome: Provides insights into the evolving role of technology in sustainability reporting.

With greater globalization and access to information, there are calls for more or less legally binding regulation of global corporate conduct and trade. Since the late 1980s, non-state actors have been involved in consolidating the moral and legal foundations of a more social and human rights-oriented global economy, including a more society and environment-oriented approach to economics. In general, a successful normative approach in the areas of corporate responsibility for human rights, labour rights and the environment are thought to be bottom-up endeavours (Ghoogassian, 2015).

One of the big actors in the promotion of sustainability reporting – and also a fighter against what it sees as social inequality devolved by economic actors – is the Global Reporting Initiative (GRI), aiming to establish comprehensive worldwide norms on the report of the economy-corporate sustainability-environment trio. This contested but promising approach to the promotion of sustainability could also be read as a profound shift in the understanding of sovereignty. The globalization of economies and communication has made the efforts of sovereign state actors to enforce social and environmental standards according to what they think is right increasingly difficult. Here, the ambition of operationalizing norms on corporate responsibility for labour rights, the environment and human rights is studied in reference to the GRI. Like other initiatives promoting sustainability reporting, the GRI recommends that disclosures and indicators should reflect overall impacts and therefore also enable stakeholders to evaluate and apply corrective measures on these impacts, likened to see also the company itself.

15.4. Challenging Questions and Answers

Q1: How does technology improve the efficiency of GRI reporting?

A1: Technology automates data collection and analysis, reduces manual errors, and enables real-time reporting, making the process faster and more accurate.

Q2: What role do digital tools play in sustainability reporting?

A2: Digital tools, such as software and analytics platforms, streamline data management, enhance transparency, and facilitate easier access to sustainability reports for stakeholders.

Q3: How can emerging technologies like AI and blockchain impact GRI reporting?

A3: AI can analyze large datasets for insights, while blockchain ensures data integrity and transparency, both enhancing the credibility and efficiency of sustainability reporting.

Q4: What challenges do organizations face when adopting technology for GRI reporting?

A4: Challenges include high implementation costs, the need for technical expertise, and ensuring data security and privacy while integrating new technologies.

Q5: How does real-time reporting benefit stakeholders?

A5: Real-time reporting provides up-to-date information, enabling stakeholders to make informed decisions and hold organizations accountable for their sustainability practices.

Q6: What is the future of technology in sustainability reporting?

A6: The future lies in advanced technologies like predictive analytics, AI, and blockchain, which will further enhance data accuracy, transparency, and stakeholder engagement.

15.5. Summary

Chapter 15 explores the transformative role of technology in GRI reporting, emphasizing its ability to enhance efficiency, accuracy, and transparency. It discusses the use of digital tools, such as software and analytics, to streamline data collection and reporting processes. The chapter also highlights emerging trends like AI and blockchain, which promise to revolutionize sustainability reporting by enabling real-time insights and ensuring data integrity. While technology offers significant benefits, challenges such as implementation costs and data security remain. Overall, the chapter underscores how technological innovation is shaping the future of sustainability reporting, making it more accessible and reliable for stakeholders.

Chapter 16
GRI and Circular Economy Reporting

Review: Explores the integration of circular economy concepts in GRI.

Key Topics: Resource efficiency, waste reduction, and lifecycle thinking.

Outcome: The chapter highlights how GRI integrates circular economy principles into sustainability reporting, emphasizing resource efficiency, waste reduction, and lifecycle thinking. It provides frameworks and indicators for organizations to measure and report on circular economy practices, fostering transparency and accountability in sustainable resource management.

16.1. Introduction

As a new paradigm for economic development, with significant environmental, economic and social benefits at the global scale, the circular economy is gaining more importance thanks to its potential to accomplish sustainable development (Berg et al., 2018). The concept of circular economy is based on the 'Three Rs' hierarchy, Reduce – Reuse – Recycle production and consumption practices by replacing the 'end-of-life' paradigm in order to close the loops. It highlights substitution of products and materials by the ones that are either renewable or biodegradable, or produced with minimal resources and contaminants. This relates to the main notion of the circular economy called 'circularity'. Promoting circularity aims to recover product and material life-cycles in order to transition to the zero-waste strategy and maintain the value of products and materials at the highest possible level. Material recirculation would in consequence decrease the need for mining, drilling and exploitation of raw resources. Creating products with a longer life-cycle due to higher durability and remanufacturing would decelerate the absolute renewable material usage and have a lower environmental impact. Adaptive regulation concerning products would impose the separation of materials, easier repair and separation processes, and tougher punitive measures in accordance with 'the polluter pays' principle. The circular economy approach will co-evolve with instant access to information about the quality, quantity and location of products and materials, which will increase the efficiency of their use inside the economy and decrease long-term irreparable sustainability failure risks. The achievement of the circular economy concept also aims at a more encompassing socio-economic development with a long-term perspective. The circular economy is not synonymous with classic GDP growth; the goal of such a concept is not to continue the industrial revolution unabated, but more to look for the most effective way to maximize well-being within the available resources and assets. Circular economy productivity is directed towards the preservation and development of human, social and natural capital. It is assumed that long-term spatiotemporal diversity will be closer to sustainable development conditions with the circular economy, which in turn can provide more equal socio-economic circumstances. Promoting the circular economy approach may lead to the reduction of disparities in capital and results might be achieved toward the convergence of the mentioned diversity to predefined boundaries. At European Union level, a number of strategies, policies, actions and regional development programs are being implemented to promote the idea of the circular economy and are largely based on existing world principles.

16.2. Circular Economy Concepts in GRI

Review: Introduces circular economy principles in GRI Standards.

Key Topics: Indicators, metrics, and reporting frameworks.

Outcome: Provides a foundation for reporting on circular economy practices. The aim of this background report is to provide a brief information package on circular economy; the concept, its history and potential, business opportunities, management, and measurement. It is argued that the shift to a circular economy offers significant diverse business opportunities and a potential increase in the circulation of materials within the economy as a way to reduce poverty. However, the significant structural changes required for this will also affect and disrupt the whole current system, and will not be implemented without substantial change from existing regulations.

The report consists of a general overview on circular economy including the origins, visions, concepts, action plans, performance measurement, businesses, bioeconomy and future, followed by a review of 16 selected articles on the economics and benefits of circular economy. This general background covers the principles and goals of circular economy, the reasons why it is considered important, the potentialities and opportunities given by a circular economy strategy, and the main future expected developments (Berg et al., 2018).

16.3. Circular Economy Performance Indicators

Review: Explores specific indicators for circular economy reporting.

Key Topics: Recycling rates, material use, and innovation.

Outcome: Helps organizations measure and report on circular economy performance.

Companies nowadays are required to continuously prove they are sustainable. Global reporting initiatives and sustainability indices demand them to disclose sustainability information, often by means of sustainability reports. Over the past decades, the market has offered a wide range of tools to produce such reports. The Global Reporting Initiative framework stands as the most widely used. A new platform, booming in the software market, claims to help organizations find the sustainable decision for their products by relying on the Global Reporting Initiative and Life Cycle Assessment metrics. On one hand, these tools play a role in disseminating sustainability assessment; but on the other, they focus narrowly on environmental concerns and may miss transformative thinking. Building on the theory of calculative agency and The Sustainable Scale Project, Life Cycle Management Platform is empirically studied. On one hand, how these tools interfere in the production of circular products (knowledge production); on the other hand, how the alternative sustainability assessment (that targets social and environmental aspects) is shaped by such tools (knowledge production) (BEGHETTO et al., 2017).

The theory of calculative agency addresses how specific tools can support certain types of actions: they make some activities possible, some more likely, others more disciplined and some prevented. As expected, this tool interferes in the production of sustainability information by 'disciplining' action. Interviews revealed how Life Cycle Management Platform is often the only way to produce a sustainability assessment in the first place. Participation in the testing phase of the platform allowed to unearth some steering effects of the tool in this context. Given an initial Product Category Rule, the set of rules product manufacturers need to follow to produce an Environmental Product Declaration, the platform steers action (i.e., asks). To some extent, this has led participants in this exchange to look for further sustainability assessment. In some cases, expectations for a broader range of social impacts may follow. On the other hand, knowing this feedback, the company develops some 'boomerang' effects (i.e., expects). Interviewees shared that production is likely to focus on continuing refining what the tool already supports, and in producing responses to the questions asked by it with an eye on achieving the highest score possible (plans) (Moraga et al., 2019).

16.4. Challenging Questions and Answers

Q1: How does GRI support circular economy reporting?

A1: GRI provides specific indicators and frameworks for reporting on circular economy practices, such as recycling rates, material efficiency, and innovation, enabling organizations to track and disclose their progress.

Q2: What are the key principles of the circular economy?

A2: The circular economy is based on the "Three Rs" hierarchy: Reduce, Reuse, and Recycle. It aims to minimize waste, extend product lifecycles, and maximize resource efficiency.

Q3: What challenges do organizations face in adopting circular economy practices?

A3: Challenges include structural changes in production processes, regulatory adjustments, and the need for innovation in material use and product design.

Q4: How do circular economy indicators benefit stakeholders?

A4: Indicators provide transparency, enabling stakeholders to assess an organization's resource efficiency, waste management, and overall sustainability performance.

Q5: What role does lifecycle thinking play in circular economy reporting?

A5: Lifecycle thinking ensures that organizations consider the environmental impact of products from creation to disposal, promoting sustainable design and resource management.

Q6: How can GRI reporting drive innovation in the circular economy?

A6: By setting clear metrics and encouraging transparency, GRI reporting motivates organizations to innovate in areas like material reuse, recycling, and sustainable product design.

16.5. Summary

Chapter 16 explores the integration of circular economy principles into GRI reporting, focusing on resource efficiency, waste reduction, and lifecycle thinking. It introduces frameworks and indicators for measuring circular economy performance, such as recycling rates and material use. The chapter highlights the benefits of circular economy practices, including reduced environmental impact and enhanced resource efficiency, while addressing challenges like regulatory changes and innovation requirements. By providing tools for transparent reporting, GRI helps organizations align with circular economy goals, fostering sustainable development and accountability. The chapter also emphasizes the role of lifecycle thinking in promoting sustainable product design and resource management, ultimately driving innovation and stakeholder trust.

Chapter 17
GRI and Supply Chain Reporting

Review: Focuses on transparency in supply chain reporting.

Key Topics: Supplier engagement, risk management, and traceability.

Outcome: The chapter emphasizes the importance of supply chain transparency in GRI reporting, highlighting how it enhances accountability, risk management, and stakeholder trust. It explores strategies for supplier engagement and performance tracking, demonstrating how GRI guidelines promote sustainable supply chain practices globally.

17.1. Introduction

It examines country-level differences in the content of stand-alone reports on sustainability according to the guidelines. Measures are applied that assess the quality of the reports' disclosure on economic, environmental, and social sustainability, and results compared between GRI-based reports of companies headquartered in emerging and developing GRI-member and non-member countries. Using a sample of 415 companies from 28 countries, differences are found between some of these on all three sustainability dimensions. The findings underline the importance of GRI membership in terms of creating the necessary internal capacities and regulatory pressures, and they highlight the limitations in the régime's scope of effectiveness and appropriateness (Barkemeyer et al., 2015).

The effectiveness of sustainability reporting reliant on the guidelines for fostering corporate social and environmental responsibility is still an open question. This study presents the findings of a multilevel analysis of 8912 companies in 42 developed and developing countries. Rather than focusing on the mere presence of GRI-based reports, the study also captures factors geared towards its regulation and normative development. It finds those GRI-based reports are more likely in developed countries and in sectors subject to greater scrutiny. A positive correlation is found between GRI-based reporting and the quality of disclosures on labour practices, whilst a GRI association is detrimental to disclosures on market place responsibility. Providing information on environmental performance is also less likely in a GRI context.

17.2. Supply Chain Transparency in GRI Reporting

Review: Explores how GRI addresses supply chain transparency.

Key Topics: Disclosure, accountability, and stakeholder trust.

Outcome: Demonstrates the value of transparent supply chain reporting.

Global Reporting Initiative (GRI) was born out of a need for consistency and comparability when, in 1997, the United States Coalition for Environmentally Responsible Economies (CERES) invited companies and environmental groups from six industrialized countries to participate in an effort to harmonize ten of the best reporting guidelines available at the time. Around this time, a small group of participants started a parallel coordinated effort with the ambition of developing a single comprehensive Sustainability Reporting Guideline. At the same time, globalization brought a wide range of global shifts that have boosted interest in Sustainable Development and lead to increased recognition of the interlinks among economic, social, and environmental issues. Furthermore, global changes resulted in a general decrease of traditional state capacities and, as a consequence, increased roles of the society and of companies as the main actors (Barkemeyer et al., 2015). Supply chain transparency according to the Global Reporting Initiative (GRI) refers to the manner in which operations and suppliers are selected, monitored, and reviewed. Best practices in supply chain transparency encourage the reporting entity and suppliers to adhere to supply chain standards and regulations, and disclose supply chain policies and performance. The effectiveness of such practices ultimately leads to better management of supply chain risks and a more sustainable supply chain. Supply chain transparency guidelines developed by the GRI are maintained in the "Social: Human Rights" aspect. Assessing disclosures made according to these GRI guidelines provides evidence on the effectiveness of private transnational governance regimes in influencing company reporting behavior.

17.3. Supplier Engagement and Reporting

Review: Discusses strategies for engaging suppliers in reporting.

Key Topics: Collaboration, standards, and performance tracking.

Outcome: Provides guidance for effective supplier engagement in GRI reporting.

In the past two decades, sustainable supply chain (SC) management has become an area of increased focus for policy makers, organizations and academics, reflected in heightened research and policy activities and the widespread incorporation of sustainability objectives further down the SC tiers (Barkemeyer et al., 2015). Holding suppliers responsible for adhering to ethical, social and environmental standards has substantial potential to promote desirable practices worldwide, including in regions where there is limited tradition of corporate social responsibility. Over the years, corporations and other organizations have been called on to take their responsibility with regard to sustainability seriously, taking action in areas such as climate change, human rights, corruption, and social welfare. In connection to this, there has been a growing public and policy interest in engaging downstream suppliers within multi-tier supply chains to adhere to similar standards. It has been argued that many traditional government processes are ill-equipped to effectively address persistent sustainability challenges, particularly due to their national focus, lack of integration and stakeholder inclusiveness, or high adaptation barriers.

Increasing supplier engagement with regard to sustainability objectives is a means to improve environmental performance and social responsibility further down the chain, maximizing the broader reach of sustainable performance. The concept of sustainable SC management has become widely acknowledged as a means to rise above risk management and cost savings, and as a way to encourage innovation, develop beneficial products and garner goodwill. Supplier engagement has been portrayed as one of the top methods to intensify SC sustainability, thereby building up partner interest in long-term progress. Efforts in various economies, including those that are less developed, have been witnessed to promote Corporate Social Responsibility (CSR) activities across the SC, with a considerable note on supplier management. Where public regulation and enforcement can be inconsistent, stakeholders are counted on to press organizations to take responsibility for the social and environmental impacts of their operations. A number of transnational voluntary initiatives have been created to guide sustainability performance for companies originating from less GRI-reporting countries. Given the many governance gaps present in these areas, adhering to the GRI seems to be a way to rebound. The GRI has managed to influence the form and content of sustainability performance in such countries.

17.4. Challenging Questions and Answers

Q1: Why is supply chain transparency important in GRI reporting?

A1: Supply chain transparency ensures accountability, reduces risks, and builds stakeholder trust by disclosing supplier practices and adherence to sustainability standards.

Q2: How does GRI promote sustainable supply chain management?

A2: GRI provides guidelines for supplier engagement, performance tracking, and disclosure, encouraging organizations to adopt ethical, social, and environmental standards.

Q3: What challenges do organizations face in supply chain reporting?

A3: Challenges include monitoring multi-tier supply chains, ensuring supplier compliance, and addressing inconsistencies in global regulations and enforcement.

Q4: How does supplier engagement improve sustainability outcomes?

A4: Engaging suppliers fosters collaboration, encourages innovation, and extends sustainability practices further down the supply chain, enhancing overall environmental and social performance.

Q5: What role do stakeholders play in promoting supply chain transparency?

A5: Stakeholders pressure organizations to disclose supply chain practices, ensuring accountability and driving improvements in sustainability performance.

Q6: How does GRI reporting address governance gaps in supply chains?

A6: GRI reporting provides a standardized framework for transparency and accountability, helping organizations in less regulated regions adopt sustainable practices and improve supply chain governance.

17.5. Summary

Chapter 17 focuses on the role of GRI in promoting supply chain transparency and sustainability. It highlights the importance of disclosing supplier practices, managing risks, and building stakeholder trust through transparent reporting. The chapter explores strategies for engaging suppliers, tracking performance, and adhering to ethical and environmental standards. It addresses challenges such as monitoring multi-tier supply chains and ensuring compliance in regions with weak regulations. By providing guidelines for supplier engagement and performance tracking, GRI helps organizations improve sustainability outcomes and address governance gaps. The chapter underscores the value of stakeholder pressure in driving accountability and fostering sustainable supply chain practices globally.

Chapter 18
GRI and Biodiversity Reporting

Review: Examines the integration of biodiversity in GRI Standards.

Key Topics: Impact assessment, conservation, and disclosure.

Outcome: The chapter highlights GRI's role in integrating biodiversity into sustainability reporting, emphasizing impact assessment, conservation efforts, and disclosure. It provides tools and frameworks for organizations to measure and report on biodiversity impacts, promoting transparency and accountability in addressing biodiversity loss.

18.1. Introduction

Introduction: Biodiversity is increasingly threatened on Earth and as a result is gaining importance on political and corporate agendas. Many organizations are causing—and mitigating—the biodiversity crisis. This sectoral focus of biodiversity research and conservation action is consistent with an understanding that the economy is both a driver of environmental change and a source of potential solutions. There is a clear need to influence the actions of organizations across the economy to benefit biodiversity, but changing corporate behavior is difficult. Biodiversity performance accountability is increased through corporate non-financial reporting, but many companies that have a large impact on biodiversity are not reporting all or relevant aspects of their impacts or biodiversity.

To encourage companies to report more about their impacts on biodiversity, and their efforts to mitigate those impacts and benefits biodiversity, conservation scientists need to engage with corporate accountability and transparency mechanisms as they relate to biodiversity. It is suggested that conservation scientists provide positive nudges to companies to encourage them to say more about relevant biodiversity impacts and actions. Although corporate accountability and transparency mechanisms may be focused on only some impacts or biodiversity, and may vary depending on the company's industry or territory, engagement with these apparent benefits the biodiversity and policy outcomes of companies generally. This work is focused on biodiversity accountability of companies that operate particularly in the forests and/or global agricultural sectors.

18.2. Integration of Biodiversity in GRI Standards

Review: Explores biodiversity-related indicators in GRI.

Key Topics: Metrics, reporting frameworks, and best practices.

Outcome: Provides tools for organizations to report on biodiversity impacts.

Based on the current Global Reporting Initiative (GRI), reporting guidelines for corporate sustainability do not contain a standard of biodiversity-related information disclosure (Mansoor, 2016). However, GRI has designed a Biodiversity Working Group to evaluate the Biodiversity Working Paper and 9 "standard draft disclosures," explored draft Implementing Measures, and measures and prompting questions. This paper's evaluation of reported biodiversity information, and disclosed information about how it should be seen within the GRI context, is, at the day's close, only the observations and views of the researcher. Biodiversity is one of the key environmental considerations in the 17th sustainable development objective. Convention on Biological Diversity lists biodiversity as "the variability among living organisms from all classifications, including, inter alia, the terrestrial, marine and other aquatic ecosystems and the ecological complexes of which they are a part; they embody diversity within species and between species; and the variety of ecosystems within every state and geographical ecosystem boundary." Changes in agricultural practices and land use, habitat clearing, and urbanization, among other things, are accelerating the loss of local and the collapse of ecosystems worldwide, and the global condition of biodiversity continues to deteriorate. On July 1, the Korean government supported the designation of the International Day for Biological Diversity. Korea's Biodiversity Use and Distribution Charge System was announced at the same opportunity. This demand device involves the collection of obligations based on the use of biogenetic properties associated to Korean species and connected to customary knowledge that will be used for economic goals, including biogenetic resources and digital information. This solemn duty applies to biogenetic resources and digital information gathered after July 1, 2008. As of July 2011, similar charges are levied on biogenetic resources and digital information for microorganisms and biogenetic resources and digital information for marine materials.

18.3. Measuring Biodiversity Impact

Review: Discusses methods for measuring biodiversity impact.

Key Topics: Assessment tools, data collection, and reporting.

Outcome: Highlights the importance of accurate biodiversity measurement in GRI.

Biodiversity is the variety of life on Earth, the natural patterns it forms and the processes that enable it to evolve, persist, and survive. Biodiversity comprises various types and scales of diversity, ranging from different species and genetics to ecosystems and biomes. Biodiversity underpins ecosystem services, which benefit humans and support food, water, energy, and health (Addison & Bull, 2018). Consideration of biodiversity and associated ecosystem services is integrated into the key elements of the Global Reporting Initiative guidelines on economic, environmental, and social sustainability.

To help protect forests, HP set a goal in 2016 to achieve zero deforestation associated with HP brand paper and paper-based product packaging. This goal is meant to have a cascading effect along the entire market. This approach shall contribute to reducing greenhouse gas emissions, human rights abuses, and challenges associated with sustainable forest management in the pulp and paper industry. During the first year, 49 companies that mentioned biodiversity and an additional 16 that mentioned sustainable forestry or fishing disclosed a range of biodiversity-related activities. These activities included managing or preventing impacts, protecting and restoring biodiversity, monitoring biodiversity, engaging and connecting people with biodiversity, and investing in biodiversity. Only 9 companies provided quantitative information about their activities, most of which was in the form of performance indicators associated with descriptions. The lack of widely used, standardized, quantitative biodiversity performance indicators create challenges for comparing performance both within industries and, particularly, between companies. The most commonly disclosed qualitative information about biodiversity activities concerned habitats protected or restored, and partnerships formed.

18.4. Challenging Questions and Answers

Q1: Why is biodiversity reporting important for organizations?

A1: Biodiversity reporting helps organizations disclose their environmental impacts, demonstrate conservation efforts, and align with global sustainability goals, enhancing transparency and accountability.

Q2: How does GRI support biodiversity reporting?

A2: GRI provides frameworks and indicators for measuring biodiversity impacts, enabling organizations to report on conservation efforts, habitat protection, and ecosystem restoration.

Q3: What challenges do organizations face in biodiversity reporting?

A3: Challenges include the lack of standardized quantitative indicators, difficulty in measuring biodiversity impacts, and ensuring comprehensive disclosure of conservation activities.

Q4: What are the key components of biodiversity impact assessment?

A4: Key components include habitat protection, species conservation, ecosystem restoration, and monitoring biodiversity changes, all of which are essential for accurate reporting.

Q5: How can organizations improve their biodiversity reporting?

A5: Organizations can adopt standardized metrics, engage in partnerships for conservation, and provide quantitative data alongside qualitative descriptions of their biodiversity efforts.

Q6: What role does biodiversity play in achieving sustainable development goals?

A6: Biodiversity underpins ecosystem services essential for human well-being, such as food, water, and health, making it critical for achieving SDG 15 (Life on Land) and other related goals.

18.5. Summary

Chapter 18 explores the integration of biodiversity into GRI Standards, emphasizing the importance of impact assessment, conservation, and transparent disclosure. It provides tools and frameworks for organizations to measure and report on biodiversity impacts, addressing challenges such as the lack of standardized metrics and comprehensive data. The chapter highlights the role of biodiversity in supporting ecosystem services and achieving sustainable development goals. By adopting GRI guidelines, organizations can enhance their biodiversity reporting, demonstrating their commitment to conservation and sustainability. Case studies illustrate successful biodiversity reporting practices, showcasing efforts in habitat protection, species conservation, and ecosystem restoration. The chapter underscores the need for accurate measurement and transparent disclosure to address global biodiversity loss effectively.

Chapter 19
GRI and Water Reporting

Review: Focuses on water stewardship in GRI reporting.

Key Topics: Water use, efficiency, and conservation.

Outcome: The chapter highlights GRI's role in water stewardship, offering frameworks for reporting on water use, efficiency, and conservation. It enables organizations to address risks, improve practices, and align with global sustainability goals. Despite GRI and ISO 14046 standards, few companies, like Lagos Water Corporation, have advanced in water reporting. Broader adoption is needed to tackle water scarcity and ensure sustainable use.

19.1 Introduction

Financial Times published an article about Strategic Water. An article was prompted by a rising awareness about human effect on natural environments and the scarcity of water supplies around the world. Key corporations claim that they are managing water supplies sustainably. The article argued that without efficient and proficient management programs, history has proven that predictions are of unprecedented water scarcity (Australia and New Zealand et al., 2013). One way to address the risks and manage it more sustainably is the publication of an ISO 14046 standard. It appears evident that only a few key companies are engaged in this sector or control such enterprises traditionally from a finance perspective. The involvement of state and non-governmental organisations has shown that companies still have a long way to go to improve disclosure and performance. Apart from improving pure water supply management and risk assessment, the GRI has presented since 2013 unique 20 water-related indicatives and one of these can be found in a table below. Based on the analysis of the maturity of the world's largest water companies, only a few have begun to report any of these and they weren't the global water and wastewater giants that you might expect. One of the key players, however, is Lagos Water Corporation, the majority Slovenian- and partly Spanish-owned Nigerian company where an active 200% return on investment was made by already the first year.

19.2 Water Stewardship in GRI Reporting

Review: Explores water stewardship principles in GRI.

Key Topics: Indicators, disclosure, and best practices.

Outcome: Provides a framework for reporting on water stewardship.

Water stewardship represents an intensification of the internal water management of organizations and a broader understanding of their role in managing water resources. Water stewardship requires the assessment of direct and indirect water use, the latter of which can be significantly larger than facility water use even for companies in water-intensive sectors (Australia and New Zealand et al., 2013). To fulfill the global needs in water management, focus on direct water use in the organization and its supply chain. It demands reporting on water use efficiency and on actual uses of water in places that are water-scarce, and areas that have competition with ecosystems. The water stewardship standard of the Water Stewardship Initiative includes the requirement to report on general water use, water sources, quality and quantity of wastewater, investment in improving water efficiency, measures to avoid contamination of water sources, water-related impact assessment, and participation in local water management dialog.

Since the introduction of sustainability reporting guidelines by the Global Reporting Initiative, topics which organizations should report on have been limited by the scope of the available guidelines (Wuerzinger, 2015). Following the release of the GRI G4 Guidelines, an analysis of five sectors revealed that water use is commonly reported since these guidelines make it obligatory. Overall, the GRI Guidelines set a minimum requirement of what organizations must report, and there are a number of material topics of sustainability reporting which are not covered by the Guidelines. Despite the benefits of GRI Guidelines, focus on reporting requirements may constrain the organizations in exploring material aspects of sustainability in better depth.

19.3. Water Use Efficiency Metrics

Review: Discusses metrics for measuring water use efficiency.

Key Topics: Consumption, recycling, and reduction strategies.

Outcome: Helps organizations track and report on water efficiency.

Regarding water discharge, amount of water recycled and reused. Disclosure was recognised for better communication of information on water as a business asset and business risk, so that the stakeholders would have the understanding necessary for investment decisions. It was for the long-term protection of natural capital, and because companies increasingly faced water scarcity in their business operations – poor quality of available fresh water, decreasing global freshwater supplies and rising demand leading to higher water and waste water disposal costs.

Disclosures on water were also seen as affecting share price, and it was suggested that Price Earnings and Dividend per Share ratios fell if bad publicity concerning a company's water use associated it with any link to pollution or inefficient use. Considered to be of most importance for the business risk disclosures were the efficient use and recycling of water. Efficiency was considered to cover two aspects of water management. The first was the minimisation of usage, especially where usage was associated with water initiatives that represented a physical intensity deadweight in the absence of a corresponding business relationship to the water used. The second aspect of efficiency was the proper need. Companies may well be involved in potentially water intensive activities; efficiency referred to the proportionality of usage to the norms of the sector.

However, instead of providing water discharge, net water efficiency, or recycling of water, only a list of water management practices was given. Management practices did not explain the efficiency or success of these initiatives; this could be through the provision of a comparison to absolute water use data or percentage claims. Furthermore, the information lacked data on productivity and left the impression that only water conservation initiatives had been taken.

19.4 Challenging Questions and Answers

Q1: What are the key challenges organizations face in adopting GRI guidelines for water reporting, and how can these challenges be addressed?

A1: Organizations often struggle with the complexity of GRI guidelines, limited resources for data collection, and the lack of expertise in water stewardship. Additionally, indirect water use in supply chains is often overlooked. These challenges can be addressed by providing training, simplifying reporting processes, and encouraging collaboration between organizations, governments, and NGOs to share best practices and resources.

Q2: How does water stewardship differ from traditional water management, and why is it critical for organizations to adopt stewardship principles?

A2: Water stewardship goes beyond internal water management by considering the broader impact of an organization's water use on ecosystems and communities. It involves assessing direct and indirect water use, engaging in local water management dialogues, and addressing water scarcity risks. Adopting stewardship principles is critical because it ensures sustainable water use, mitigates business risks, and enhances an organization's reputation and stakeholder trust.

Q3: Why is the disclosure of water efficiency metrics important for stakeholders, and how can organizations improve their reporting in this area?

A3: Disclosure of water efficiency metrics is crucial because it provides stakeholders with insights into an organization's water management practices, risks, and sustainability performance. Improved reporting can be achieved by providing quantitative data (e.g., water recycled, net water efficiency) and benchmarking against industry norms. Organizations should also highlight specific initiatives and their outcomes to demonstrate progress.

Q4: What role do ISO 14046 and GRI play in addressing global water scarcity, and how can their adoption be encouraged?

A4: ISO 14046 and GRI provide standardized frameworks for assessing and reporting water use, efficiency, and risks. They help organizations identify areas for improvement and align their practices with global sustainability goals. Adoption can be encouraged through regulatory incentives, stakeholder pressure, and demonstrating the financial and reputational benefits of compliance.

Q5: How can organizations balance the need for detailed water reporting with the potential constraints imposed by GRI guidelines?

A5: Organizations can balance this by going beyond the minimum GRI requirements and exploring material aspects of water stewardship in greater depth. They should also integrate water reporting into their overall sustainability strategy, ensuring it aligns with business objectives and stakeholder expectations. Collaboration with industry peers and experts can help identify innovative reporting practices.

Q6: What lessons can be learned from the Lagos Water Corporation's success in water stewardship, and how can other organizations replicate this success?

A6: Lagos Water Corporation's success highlights the importance of setting clear goals, investing in water efficiency initiatives, and engaging stakeholders. Other organizations can replicate this by adopting similar strategies, leveraging partnerships, and using frameworks like GRI and ISO 14046 to guide their efforts. Transparency and accountability in reporting are also key to building stakeholder trust.

19.5 Summary

This chapter highlights the growing awareness of water scarcity and the need for sustainable water management. It discusses the role of ISO 14046 and GRI standards in improving water stewardship, emphasizing metrics like water use efficiency, recycling, and risk disclosure. While some companies, like Lagos Water Corporation, show progress, many lag in reporting. The chapter underscores the importance of transparent water management for business risk, stakeholder trust, and long-term sustainability, while noting gaps in current reporting practices.

Chapter 20
GRI and Anti-Corruption Reporting

Review: Examines the role of GRI in anti-corruption reporting.

Key Topics: Transparency, accountability, and ethical practices.

Outcome: The chapter highlights GRI's role in promoting anti-corruption measures through transparent and accountable reporting. It provides tools and frameworks for organizations to disclose anti-corruption efforts, enhancing corporate integrity and reducing corruption risks. Case studies demonstrate the effectiveness of GRI in improving transparency and reducing corruption-related incidents.

20.1. Introduction

It is investigated whether companies in countries with poor quality of legal institutions tend to produce low quality corporate sustainability reports with significantly lower anti-corruption reporting compared to companies in countries with high-quality legal institutions. Content analysis is conducted on a sample of sustainability reports published between 2002 and 2012 by companies participating in the GRI. The likelihood of anti-corruption reporting and the use of spun information are tested. For the reason of limited resources, there is chosen not to examine all of the GRI performance indicators related to anti-corruption. Treatment of stakeholders as part of the governance structure of organizations is emphasised as an increasing concern worldwide. There is an increase in the attention given to anti-corruption issues by both the governments of the home countries of TNCs and at the international level. Five years after the release of the UN Global Compact, a survey finds that a significant percentage of companies from various countries had already expressed their support for the initiative. Furthermore, a notable percentage of respondents pointed to the Global Reporting Initiative as one of the tools used by companies in order to increase transparency in their activities. Generally, a substantial percentage of the world's largest transnational corporations claim adherence to the GRI guidelines. This initiative has set the present standard for corporate social reports. Nevertheless, anti-corruption reporting is relatively new and different factors may hamper countries in writing about their programs and activities in this field. This dependence arises from the information published and the overall quality of reports. Different sectors took advantage of the possibility to write partially on certain issues and completely escape from writing on others. In particular, there is no company reporting even one time in five on compliance, checks, corruption-related complaints and investigations. Furthermore, the disclosure remains superficial on many other important issues.

20.2. Incorporating Anti-Corruption Measures in GRI Reporting

Review: Explores how GRI addresses anti-corruption in its standards.

Key Topics: Indicators, disclosure, and best practices.

Outcome: Provides tools for organizations to report on anti-corruption efforts.

According to the Global Corruption Barometer (Barkemeyer et al., 2015), among the general public in 107 countries, 27% of all documents, alongside other resources and facilities, altered their Australian legislation or organizations in support of business operations. 16% paid bribes or used influential connections in order to succeed in Australian public and supplier procurement tender processes. Compliance-oriented companies "are then likely to require disclosure, in [corresponding] sustainability reports, of the mechanisms in place to prevent corruption." Consequently, the Global Reporting Initiative (GRI) incorporated anti-corruption-related indicators into its voluntary guidelines between 2000 and 2013, setting in its 2000 guidelines a requirement for organizations to disclose any "company's public policy and organizational mechanisms to combat corruption." In the period between 2002 and 2012, adherence to the GRI reporting was substantially better in emerging market multinationals. In order to increase globalized economy transparency, the Australian government, in the Asia-Pacific Economic Cooperation (APEC) leaders' summit in Bogor, Indonesia, adopted the Asia-Pacific Partnership by endorsing the APEC Transparency Standards. These standards are defined by the APEC as "voluntary, based on existing APEC agreements, and should be practical and achievable and flexible, taking into account different levels of development and legal systems, while encouraging sharing of best practices" (Ghoogassian, 2015). Since the formation of the APEC in 1989, the dialogue has failed to assess trade compliance of its 21 members. The full implementation of these transparency standards is not enforced as they are legally non-binding on the APEC member economies, and also because officials involved in corruption would face public disclosure as well as potential legal action.

20.3. Case Studies on Anti-Corruption Reporting

Review: Presents examples of anti-corruption reporting in practice.

Key Topics: Success stories, challenges, and outcomes.

Outcome: Illustrates effective anti-corruption reporting using GRI.

Multinational Enterprises (MNEs) have exacerbated corruption in many ways, such as providing improper inducements or gifts perceived as bribery by the enticement laws of the foreign country in which they are provided, forming cartels under the guise of restructuring or reorganizing in order to maximize shareholders' profits, using an affiliate account to commit corruption, and investing in countries whose governments are known for corruption. (Wu et al., 2017) The findings provide robust evidence that the voluntary adoption of the GRI has a material and persistent effect on both disclosure transparency and the quality of disclosed anti-corruption efforts. Without the GRI, MNEs that operate in foreign environments known for corruption start to act in a less corrupt way after being named in the reports. On the other hand, with the GRI, MNEs tend to respond by 'engaging in less corruption', and a lower number are named in composite investigations. Consequently, 10% fewer GRI-reporting MNEs are implicated in anti-corruption investigations compared to non-GRI-disclosing MNEs. Similarly, the informal announcement using the GRI exacerbates the financial losses associated with such investigations by a factor of almost 1000 nature partners.

20.4. Challenging Questions and Answers

Q1: How does GRI support anti-corruption reporting?

A1: GRI provides specific indicators and guidelines for disclosing anti-corruption measures, enabling organizations to report on policies, mechanisms, and compliance efforts to combat corruption.

Q2: What are the key challenges in anti-corruption reporting?

A2: Challenges include limited resources, superficial disclosures, and varying legal and institutional frameworks across countries, which can hinder comprehensive reporting.

Q3: How does anti-corruption reporting benefit organizations?

A3: Anti-corruption reporting enhances corporate transparency, builds stakeholder trust, reduces legal and reputational risks, and aligns with global ethical standards.

Q4: What role do stakeholders play in promoting anti-corruption measures?

A4: Stakeholders, including governments, investors, and the public, pressure organizations to adopt transparent practices and disclose anti-corruption efforts, driving accountability.

Q5: How effective is GRI in reducing corruption in multinational enterprises (MNEs)?

A5: GRI reporting has been shown to reduce corruption incidents in MNEs, with fewer GRI-reporting companies implicated in anti-corruption investigations compared to non-reporting ones.

Q6: What are the best practices for anti-corruption reporting?

A6: Best practices include disclosing anti-corruption policies, mechanisms for compliance, whistleblower protections, and quantitative data on corruption-related incidents and investigations.

20.5. Summary

Chapter 20 examines GRI's role in anti-corruption reporting, emphasizing transparency, accountability, and ethical practices. It provides frameworks and indicators for organizations to disclose anti-corruption measures, addressing challenges like limited resources and varying legal frameworks. The chapter highlights the benefits of anti-corruption reporting, including enhanced corporate integrity and reduced risks. Case studies demonstrate GRI's effectiveness in reducing corruption incidents, particularly in multinational enterprises. By adopting GRI guidelines, organizations can improve transparency, build stakeholder trust, and align with global ethical standards, contributing to a more accountable and ethical business environment.

Chapter 21
GRI and Circular Reporting

Review: Explores the concept of circular reporting in GRI.

Key Topics: Principles, frameworks, and examples.

Outcome: The chapter emphasizes the importance of circular reporting in GRI, focusing on lifecycle thinking, resource efficiency, and innovation. It provides frameworks and real-world examples to demonstrate how circular reporting enhances sustainability practices, offering detailed and in-depth insights into economic, environmental, and social performance.

21.1. Introduction

Global sustainability concerns have led, over the last two decades, to the setting up of non-binding platforms for public, voluntary sustainability reporting. First in 1997, and thereafter in 2000 and 2013, the Global Reporting Initiative (GRI) established voluntary guidelines and indicators for economic, environmental and social (EES) performance. Among their various objectives, GRI intended at best to enhance EES accountability and transparency, increasing the comparability, reliability and relevance of disclosed information. Controversy around the GRI standards has usually pertained to their impact on both the scope and the quality of EES disclosure. The aim of the presented study is to investigate the adoption and impact of GRI standards, focusing on Italian non-financial listed companies and introducing a domain-specific circular disclosure index. Indeed, accounting for the specific nature of reporting topics, a dedicated circular disclosure index is developed to measure the extent to which companies pair GRI recommendations and indicators, creating more detailed and in-depth information on EES performance in terms of both strategy and implementation (Pavaloaia et al., 2017). The results indicate a positive and significant association between GRI adoption and more detailed, circular, EES disclosure, supporting the view of the framework as a guiding tool for sustainability reporting improvement.

21.2. Circular Reporting Concepts and Principles

Review: Introduces the principles of circular reporting.

Key Topics: Lifecycle thinking, resource efficiency, and innovation.

Outcome: Provides a foundation for understanding circular reporting.

Abstract: The objective of ''GRI'' is to facilitate voluntary public reporting by organizations on economic, environmental and social (sustainability) issues, to complete a consistent and globally comparable structure. The standards of ''GRI'' promote services including indicators to identify and report the outcomes on issues that are important to the way business operates. The main flexibility with the ''GRI standards'' is sector-specific and general standards may be used in combination. The document reports upon the initiative of GRI, upon its nature and the current development, as well as upon experiences gained within companies along with opinions of business organizations and other advocacy groups. ''GRI'' is an Amsterdam based international initiative, commenced in 1997, whose objective is to facilitate voluntary public reporting by organizations on economic, environmental and social (sustainability) issues. Within an emerging framework of regulation and other factors that may bear on such reporting, the initiative undertakes, with the assistance of an international network of organizations and expertise, to complete a consistent and globally comparable structure for sustainability reporting principles.

21.3. Examples of Circular Reporting

Review: Presents real-world examples of circular reporting.

Key Topics: Case studies, best practices, and outcomes.

Outcome: Demonstrates successful applications of circular reporting.

The Global Reporting Initiative (GRI) was established in 1997 as an independent organization funded jointly by the Coalition for Environmentally Responsible Economies (CERES) and the United Nations Environment Programme (UNEP) (Ghoogassian, 2015). GRI's mission is to support the widespread integration of sustainability reporting "by all organizations," which it claims will "advance sustainable development practice and performance". It is expected that companies will use the GRI Reporting Guidelines to "build their stakeholders' trust and to develop their longer-term sustainable practices". GRI defines sustainability reporting as "the practice of preparing a disclosure about an organization's economic, environmental, and social performance". GRI argues that it is "in the interest of an organization's stakeholders" that all organizations are "transparent" about their performance and that disclosures should reflect "material impacts" and should be reported in such a way as to enable stakeholders to "take decisions in response to the information disclosed".

Circular reporting around this issue is a central part of how GRI is operationalizing the emergence of norms elaborating, and in some cases supplementing, both hard and soft law on corporate responsibility for human rights. There are six examples of circular reporting. Broadly, after outlining the semi-hard legal status of sustainability reporting obligations emerging under hard and soft law, the article provides an account of GRI and its work promulgating these norms. Changing Lenses on the World was the official introduction of GRI and its guidelines to the business community and though it would be some time before it became the Reporting Guidelines that are familiar today, this initial work constituted GRI's foundational contribution to the law at the dawn of the age of sustainability reporting.

21.4. Challenging Questions and Answers

Q1: What is circular reporting in the context of GRI?

A1: Circular reporting involves integrating lifecycle thinking, resource efficiency, and innovation into sustainability reporting, providing a comprehensive view of economic, environmental, and social impacts.

Q2: How does circular reporting differ from traditional sustainability reporting?

A2: Circular reporting focuses on the entire lifecycle of products and processes, emphasizing resource efficiency and innovation, whereas traditional reporting often addresses isolated aspects of sustainability.

Q3: What are the key principles of circular reporting?

A3: Key principles include lifecycle thinking, resource efficiency, innovation, and the integration of economic, environmental, and social performance metrics.

Q4: How does GRI support circular reporting?

A4: GRI provides frameworks and indicators that guide organizations in adopting circular reporting practices, enhancing the depth and comparability of sustainability disclosures.

Q5: What are the benefits of circular reporting for organizations?

A5: Benefits include improved resource efficiency, enhanced stakeholder trust, better risk management, and alignment with global sustainability goals.

Q6: Can you provide an example of successful circular reporting?

A6: Companies like HP have adopted circular reporting by setting goals for zero deforestation in their supply chains, demonstrating lifecycle thinking and resource efficiency.

21.5. Summary

Chapter 21 explores the concept of circular reporting within the GRI framework, emphasizing lifecycle thinking, resource efficiency, and innovation. It introduces principles and frameworks that guide organizations in adopting circular reporting practices, enhancing the depth and comparability of sustainability disclosures. The chapter presents real-world examples, such as HP's zero deforestation goal, to illustrate successful applications of circular reporting. By integrating circular reporting principles, organizations can improve resource efficiency, build stakeholder trust, and align with global sustainability goals. The chapter underscores the importance of detailed and comprehensive reporting in driving sustainable development and operationalizing norms around corporate responsibility.

Chapter 22
GRI and Voluntary vs. Mandatory Reporting

Review: Compares voluntary and mandatory reporting under GRI.

Key Topics: Benefits, challenges, and trends.

Outcome: The chapter compares voluntary and mandatory reporting under GRI, highlighting the benefits of flexibility and stakeholder engagement in voluntary reporting, while emphasizing the growing trend towards mandatory reporting driven by regulatory demands. It provides insights into the evolving landscape of sustainability reporting requirements.

22.1. Introduction

The movement concerning mandatory reporting regards not only the idea of imposing a number of companies the obligation to publish a sustainability report but also the idea of establishing what an ideal report should contain. Regarding the first aspect, there are different practices at international level. In some countries, for example Austria, Czech Republic, Germany, Hungary, Italy, Norway, Poland and Portugal, big companies have the legal obligation to publish a nonfinancial report, while in other countries (i.e. the United Kingdom) are required to report pieces of information regarding corporate governance, including information on their behavior in relation to stakeholders. Contrary, Denmark, Spain, Slovenia and again Italy have extended CSR best practices independently. Because of these different practices, the European Commission tried to regulate the subject and representing a step in the direction of a unique model of CSR reporting in EU Member States, the Commission Recommendation 2001/453/EC was adopted (Pavaloaia et al., 2017). In accordance with this legal act, the report should include, besides figures and data, also a description of the company's policies, the results of those policies in terms of economic, environmental and social tools, the methodology used to determine the effects of the company's activities, etc. On the other hand, the hierarchy of findings of a future ideal report is not required by any legal act. All frameworks of reporting, respectively the Global Reporting Initiative (GRI), the European Federation of Financial Analysts Societies, the US based Global Financial Markets Association or Transparency & Accountability, focus on the connection between corporate CSR environment. These frameworks (especially the first one) contain an impressive list of elements that may be comprised in a report. Another aspect to be considered relates in drawing a theoretical framework of reporting by choosing the theories that may be utilized to explain the readiness of a company to adopt CSR and the shape of this adopted is voluntarily reporting. Last but not least, it is important to establish how the aforementioned theories reflect the content of reports, a content that can be correlated with GRI G4 indicators. The last aspect of the research is done in a quantitative manner and refers to a content analysis of CSR reports issued by big Romanian companies.

22.2. Benefits and Challenges of Voluntary Reporting

Review: Explores the pros and cons of voluntary reporting.

Key Topics: Flexibility, credibility, and stakeholder expectations.

Outcome: Offers a balanced view of voluntary reporting under GRI.

Since 1990s, international environmental, social, and governance criteria were developed for the investors as a result of the ethical consideration related to the globalization of the economy. These considerations were enhanced by the UN Global Compact introduced in 1999 which promotes a set of principles that are subject of monitoring through the submission of the so called COP reports. Between them there are social responsibility, human rights, environment and anti-corruption.

The organizations which adopted the set of principles were "playing double"; on the one hand they respected them and on the other hand they published a report for that in order to increase their image in the eyes of the customers, the suppliers, the employees, the shareholders and the state representatives. All of these referred as the stakeholders of the organization.

There are two strategies in raising the organization's image on the top resulted of the sustainable development. Between them there are generated benefits of the sustainable development and the adoption of sustainable practices (Pavaloaia et al., 2017). On their base some medium and long term objectives were established parallel with the decisions for the adoption of voluntary reporting.

22.3. Trends towards Mandatory Reporting

Review: Discusses the shift towards mandatory reporting.

Key Topics: Regulatory developments, global trends, and implications.

Outcome: Highlights the growing importance of mandatory reporting in sustainability.

In response to growing demand for sustainability data, the first initiative to create sustainability reporting guidelines was taken in 1997 by the Coalition for Environmentally Responsible Economies, the United Nations Environment Programme, and the United Nations Conference on Trade and Development. Shortly afterwards, the Global Reporting Initiative was established to develop guidelines for the sustainability reporting of economic, environmental, and social performances following a long consultative process, which also included businesses (Ghoogassian, 2015).

Today, an increasing number of governments and lawmakers are driving organizations towards becoming more sustainable by issuing guidelines, regulations, or laws that demand companies to be more transparent about their environmental and societal behavior. In the EU, for instance, the European Commission argues that "there is a need to improve the comparability of social and environmental information and to monitor the way that businesses manage social and environmental challenges." In October 2014, the EU Parliament agreed upon a directive to introduce new transparency requirements that compel large publicly-listed European companies to disclose information on their policies, risks, and results as regards environmental matters, social and employee-related aspects, respect for human rights, anti-corruption, and bribery matters, as well as activities to prevent such incidences. Such reporting obligations have to be integrated in the management report or, for some entities, in a new unique statement.

22.4. Challenging Questions and Answers

Q1: What are the key benefits of voluntary reporting under GRI?

A1: Voluntary reporting offers flexibility, enhances corporate credibility, and allows organizations to tailor disclosures to stakeholder expectations, fostering trust and engagement.

Q2: What challenges do organizations face with voluntary reporting?

A2: Challenges include inconsistent reporting quality, lack of comparability, and potential greenwashing, which can undermine credibility and stakeholder trust.

Q3: Why is there a shift towards mandatory reporting?

A3: Mandatory reporting is driven by the need for greater transparency, comparability, and accountability in addressing environmental, social, and governance (ESG) issues.

Q4: How does mandatory reporting impact organizations?

A4: Mandatory reporting increases compliance costs and administrative burdens but improves transparency, reduces risks, and aligns organizations with global sustainability standards.

Q5: What role do regulatory developments play in sustainability reporting?

A5: Regulatory developments, such as the EU's Non-Financial Reporting Directive, standardize reporting requirements, enhancing comparability and accountability across organizations.

Q6: How does GRI support both voluntary and mandatory reporting?

A6: GRI provides frameworks and guidelines that organizations can use for both voluntary and mandatory reporting, ensuring consistency, comparability, and alignment with global standards.

22.5 Summary

Chapter 22 compares voluntary and mandatory reporting under GRI, exploring the benefits and challenges of each approach. Voluntary reporting offers flexibility and stakeholder engagement but faces issues like inconsistency and greenwashing. Mandatory reporting, driven by regulatory demands, enhances transparency and comparability but increases compliance costs. The chapter highlights the growing trend towards mandatory reporting, exemplified by the EU's Non-Financial Reporting Directive, which standardizes ESG disclosures. GRI supports both approaches by providing frameworks that ensure consistency and alignment with global sustainability standards. The chapter underscores the evolving landscape of reporting requirements, emphasizing the need for transparency and accountability in addressing environmental, social, and governance challenges.

Chapter 23
GRI and Reporting in Emerging Markets

Review: Examines the adoption of GRI in developing economies.

Key Topics: Challenges, opportunities, and growth potential.

Outcome: The chapter highlights the growing adoption of GRI in emerging markets, showcasing its potential to drive sustainability reporting despite challenges like resource constraints and regulatory gaps. It emphasizes opportunities for capacity building and the role of GRI in fostering transparency and accountability in developing economies.

23.1. Introduction

An announcement was made that MBCCI, a group of local associations, will offer various services to help prepare and standardize sustainability reports in the beginning of 2013. The global reporting initiative (GRI) officially announced the content of the new G4 guidelines in May (Wuerzinger, 2015). These combined events are the perfect moment to monitor the behavior of Portuguese and Spanish organizations towards the most recent guidelines, and thus make a meta-analysis of the public's response. A recent research undertaking by GRI contracted seven researchers to contribute overviews of all the publicly available sustainability reports from selected countries with strong emerging economies and financial giants not part of the OECD, among which was included China. A substantial number of Chinese businesses heavily invest in developing countries in Africa and South America. China was not part of the previously conducted research project providing an insight into a GRI database (Romana, 2018).

23.2. Adoption of GRI in Developing Economies

Review: Explores the uptake of GRI in emerging markets.

Key Topics: Case studies, barriers, and success factors.

Outcome: Highlights the potential for GRI to drive sustainability in developing economies.

This section is devoted to the empirical analysis, which is structured as follows: variables are introduced first, followed by a description of the research design and the analytical strategy used for estimation. Then the results are presented and interpreted, and potential endogeneity issues are discussed. A systematic perspective from which emerging countries are inching their way towards sustainable development practices, and in which the GRI appears to have played a key function as the first transnational normative body in the sustainability reporting landscape. Then attention is turned on developing countries as a whole, considering them as a group of heterogeneous economies characterised in general by poor regulatory quality and low environmental sustainability performances and only recently engaged in the sustainability reporting journey. After that matters are narrowed down to give an insight into the advent of corporate sustainability reporting in emerging countries, with the empirical analysis on the adoption of the GRI framework aimed at testing two research hypotheses about information increases and the ability of the GRI to homogenise the reporting area in developing markets. A firm level empirical analysis on 1282 company reports in this particular part of corporate sustainability reporting has been conducted, in order to offset wide ranging literature review gaps that have particularly affected the developing country context analysis: in the meantime, the SAF has wrought, as of 2017, a truly confirmative experience, at least in terms of ex-ante predictions, on the empirically weak foundations of the political expectations placed in corporate sustainability reporting as a scope for inducing substantial improvement in corporate social performances of developing economies.

23.3. Challenges and Opportunities in Emerging Markets

Review: Discusses the obstacles and potential for GRI reporting in emerging markets.

Key Topics: Resource constraints, lack of awareness, regulatory gaps, and opportunities for capacity building.

Outcome: Provides a balanced view of the challenges faced by emerging markets while highlighting opportunities for growth and improvement in sustainability reporting.

The 2016 Global Reporting Initiative (GRI) Sustainability Reporting Guidelines has five major groups of performance indicators on: economic, environment, labor, human rights, and society. The framework aims for standard format and encourages the inclusion of additional information to support more detailed reports (Romana, 2018). Furthermore, the 2015 Paris Agreement has 195 signatories pledging to limit global warming to 2 degrees Celsius. The agreement promotes transparency and accountability, and climate-related financial information is encouraged in mainstream reports. The adaption of these new indicators increases the scope of corporate environmental and social responsibilities. It is thus timely and pertinent to examine the responses of MNCs headquartered in Hong Kong to these new developments. The city is an ideal testing ground as it has experienced the reintegration into China, and is an important gateway to Chinese markets. The disclosure strategies considered include by whom, when, and how much disclosure. The "by whom" issue examines whether companies with prior publication of the 2013 GRI ready-made report are more likely to publish 2016 standard compliant reports. Since reports generated by the GRI check software have standard formats and pre-determined contents meeting the guidelines, unformatted written reports prepared in-house or by consultants are considered less detailed reports. Therefore, the "how much disclosure" issue is examined in terms of the format of this extra GRI plus information included. To the best of the author's knowledge, no previous study of the 2016 guidelines focuses on Hong Kong-listed MNCs.

23.4. Challenging Questions and Answers

Q1: What are the main challenges for GRI adoption in emerging markets?

A1: Challenges include resource constraints, lack of awareness, weak regulatory frameworks, and limited technical expertise in sustainability reporting.

Q2: How does GRI support sustainability reporting in developing economies?

A2: GRI provides standardized frameworks and guidelines, helping organizations in emerging markets improve transparency, accountability, and sustainability practices.

Q3: What opportunities does GRI offer for emerging markets?

A3: GRI offers opportunities for capacity building, enhancing corporate credibility, and aligning with global sustainability standards, fostering economic and social development.

Q4: How can emerging markets overcome barriers to GRI adoption?

A4: Overcoming barriers requires investment in education, technical training, regulatory support, and partnerships with international organizations to build reporting capacity.

Q5: What role do multinational corporations (MNCs) play in promoting GRI in emerging markets?

A5: MNCs can drive GRI adoption by setting examples, sharing best practices, and encouraging local suppliers and partners to adopt sustainability reporting standards.

Q6: How does GRI reporting benefit emerging markets?

A6: GRI reporting enhances transparency, attracts investment, improves stakeholder trust, and supports sustainable development goals in emerging markets.

23.5. Summary

Chapter 23 examines the adoption of GRI in emerging markets, highlighting both challenges and opportunities. While resource constraints, lack of awareness, and regulatory gaps pose significant barriers, GRI offers frameworks that enhance transparency and accountability. The chapter emphasizes the potential for capacity building and the role of multinational corporations in driving GRI adoption. Case studies illustrate how GRI is expanding its reach in developing economies, fostering sustainability reporting and aligning with global standards. The chapter underscores the importance of GRI in promoting sustainable development, improving corporate credibility, and supporting economic growth in emerging markets. By addressing challenges and leveraging opportunities, GRI can play a pivotal role in advancing sustainability practices worldwide.

Chapter 24
GRI and Non-Financial Reporting Directive (NFRD)

Review: Explores the relationship between GRI and the EU's Non-Financial Reporting Directive (NFRD).

Key Topics: Alignment of GRI Standards with NFRD requirements and their combined impact on corporate reporting.

Outcome: The chapter highlights the alignment between GRI Standards and the EU's Non-Financial Reporting Directive (NFRD), emphasizing how GRI supports NFRD compliance. It demonstrates how GRI enhances the quality, consistency, and transparency of non-financial reporting, streamlining processes for organizations across Europe.

24.1. Introduction

The Non-Financial Reporting Directive (NFRD) of the European Union was introduced in 2014 to increase corporate transparency on environmental, social, gender and employee matters. The collected data is shared with stakeholders, enabling them to make responsible investments, economic decisions, and hold organizations accountable for their impacts. It helps identify risks, seize the opportunities, and increase resilience. The GRI Standards are global sustainability and Integrated Reporting standard that was initiated in 1997. The mission of GRI is to enable more responsible decision making by capturing, analyzing, and reporting ESG information. They are the most widely used standards for measuring and reporting sustainability. GRI releases three standards each for different organizations: the Essential, Comprehensive and the TBT Standards. The use GRI and the implementation of GRI standards increased thanks to the collaboration. The Global Resource Institute contributed to the development of the NFRD and has actively supported the better application of the NFRD since then, recognizing the many benefits of more extensive corporate reporting.

24.2. Overview of NFRD

Review: Provides an introduction to the NFRD and its objectives.

Key Topics: Scope, requirements, and the directive's focus on environmental, social, and governance (ESG) issues.

Outcome: Clarifies the purpose and significance of NFRD in promoting transparency and accountability in corporate reporting.

Founded in 1997, Non-profit organization Global Reporting Initiative (GRI) has become one of the most important reporting systems in Europe, as its guide, standards, to be fulfilled by at least 2 elements of 20, are the starting reporting data for a closer sustainable attitude of entities larger than 500 employees. Although is not compulsorily, the awareness and efforts of non-profit organizations in fulfilling them are recognized by the authorities, praised in the annual CSR awards ceremony, and exemplified in this regard. Such a report appears to be the first study on this subject and helps in a better understanding of the vision and activity of the third sectors in a sustainable economy. For revealing their involvement in the sustainable activity of society, reports on sustainable reporting were chosen, since they are more complex and offer a better picture on the particularly reported non-profit entities (Pavaloaia et al., 2017).

Designed as a tool to increase the transparency, the non-commercial approach and the unsalability of internal information, G4 guidelines of the Global Reporting Initiative are now structured on a modular approach: general definitions and contents, standard discloser items, and specific standard discloser items for each of the 46 economic sectors. Although the GRI guidelines are to be planned in a modular manner for 100 crucial economic sectors, the international comparability of the data related to the energy sector is significant and must comply with the most important environmental, social and economic issues.

24.3. Harmonization with GRI Standards

Review: Examines how GRI Standards align with and support NFRD compliance.

Key Topics: Shared principles, reporting synergies, and the benefits of harmonization.

Outcome: Demonstrates how GRI facilitates NFRD compliance, streamlining reporting processes for organizations.

Sustainability is the golden word of the 21st century. Who wants to be accused of not being sustainable? What better way to have this as a badge than to be able to prove it? The GRI standards organizations can measure and interpret how sustainably they operate. In contrast to a report according to GRI standards is something palpable. Everybody knows what to do with a report. The most important information is presented in a structured way. All relevant aspects are covered. To make this vast amount of information accessible a summary is provided. At the end there are tables that compare the key figures of the years at a glance. The GRI standards have been developed to facilitate the publication of sustainability reports (Barkemeyer et al., 2015). One of the key factors that make the GRI standards useful to this aim is that they allow the reader to distinguish between the good and the bad, and with some limitations about the ugly, or the necessary but not yet sufficiently covered. Personal judgments are kept predominantly out of the assessment. The standards offer criteria, and with few exceptions the assessment is based on these. For instance, the guidelines require information on all 'core' indicators, a carefully chosen selection of issues according to the relevance for the enterprise and its stakeholders, and additional indicators to the GRI-program of sustainability. There is no report where this method has not applied. Recommended disclosures are only referred to if information should have been published according to the guidelines but has not been found in the reports. The middle part of the table raises "common" issues or deficits (in plausibility, coverage, explanation etc.) only if at least two thirds of the reports do not adhere to the requirements. Under the hash sign specific questions and other developments are outlined.

24.4. Challenging Questions and Answers

Q1: How does GRI support compliance with the NFRD?

A1: GRI provides standardized frameworks and indicators that align with NFRD requirements, helping organizations meet reporting obligations while ensuring transparency and comparability.

Q2: What are the key objectives of the NFRD?

A2: The NFRD aims to increase corporate transparency on environmental, social, and governance (ESG) issues, enabling stakeholders to make informed decisions and hold organizations accountable.

Q3: What are the benefits of harmonizing GRI Standards with the NFRD?

A3: Harmonization simplifies reporting processes, enhances data consistency, and ensures that organizations meet both GRI and NFRD requirements efficiently.

Q4: How does GRI enhance the quality of non-financial reporting under the NFRD?

A4: GRI's comprehensive frameworks and indicators ensure that organizations provide detailed, structured, and comparable ESG data, improving the overall quality of non-financial reports.

Q5: What challenges do organizations face in aligning GRI with NFRD?

A5: Challenges include managing complex reporting requirements, ensuring data accuracy, and integrating GRI Standards with NFRD's specific disclosure obligations.

Q6: How does GRI contribute to stakeholder trust under the NFRD?

A6: By providing transparent and standardized ESG data, GRI helps organizations build stakeholder trust, demonstrating accountability and commitment to sustainability.

24.5. Summary

Chapter 24 explores the relationship between GRI Standards and the EU's Non-Financial Reporting Directive (NFRD), highlighting their alignment and synergies. The NFRD aims to enhance corporate transparency on ESG issues, while GRI provides standardized frameworks that support NFRD compliance. The chapter emphasizes how GRI streamlines reporting processes, ensuring consistency and quality in non-financial disclosures. By harmonizing GRI Standards with NFRD requirements, organizations can meet regulatory obligations while improving transparency and stakeholder trust. The chapter underscores the importance of GRI in facilitating sustainable reporting practices across Europe, contributing to greater accountability and informed decision-making.

Chapter 25
GRI and Future Trends in Reporting

Review: Looks ahead at the evolving landscape of sustainability reporting and GRI's role in shaping future trends.

Key Topics: Innovations, emerging practices, and predictions for the future of reporting.

Outcome: The chapter explores the future of sustainability reporting, emphasizing GRI's role in driving innovation and standardization. It highlights emerging practices like digital tools, AI, and blockchain, and predicts increased global adoption of GRI Standards, positioning GRI as a leader in shaping the next generation of reporting.

25.1. Introduction

Only the GRI Database formed a global common understanding of report content and organization, made reports more comparable and ensured that they were not entirely promotional. Since the sustainability report took off, reporting companies "designed" it according to their needs: they introduced various formats, different indices or internal structures, which contents were accordingly loosely organized, either inconsistent or lacking in some respect. Moreover, companies debated which issues were material and avoided, that is not prepared, other ones. The interest in an extensive common "reporting vocabulary" brought to the birth of voluntary, yet widely adopted, reporting guidelines, among which the GRI guidelines have shaped the scene for some years. On the 22nd of May 2013 the G4 guidelines were launched. They were the most significant revision of the reporting guidelines by the GRI and involved a substantial overhaul of the text and content of disclosures. In the face of such hidden problems, despite of a sustained decrease over the years, course readings aimed at "answering" about 25% of reports. While before G3 guidelines "chunks" of replies they often matched inspections by being located in similar spots within different reports, akin to interviews they were not completely "free" as not delimited by the availability of directed questions. Rather, questions more likely guided towards issue thus favoured their reporting, examples being treatment on policy for employees/staff lacking job security or measures taken to assess customer satisfaction. During the course-taking period substantial efforts were made to overcome the situation. Efforts mainly targeted within the GRI questions. While 84 questions out of 84 had been prepared, "flags" on the Research Tool were set on questions 1 to 22. Each of this 22 questions pertained to a particular issue and was treated in the same way. Specifically 3-4 replies were prearranged in order to exploit the commonalities across reports as a means to gain full marks on the issue. Another 4-5 supporting issues were "checked" off in the hope that, considering the somewhat indirect presence of some information on a report, they could falsely enhance the successful treatment of any of the 22 core issues high (or maximum) level of full marks.

25.2. Innovations in Reporting Practices

Review: Explores new and emerging practices in sustainability reporting influenced by GRI.

Key Topics: Digital tools, real-time reporting, and integration of advanced technologies like AI and blockchain.

Outcome: Shows how GRI is leveraging innovation to enhance the accuracy, efficiency, and impact of sustainability reporting.

To provide a new way for companies to increase their accountability to the wider public regarding their sustainability initiatives, a reporting initiative was founded in Boston, USA in 1997. The obligation to report applies only to very large corporations with the number of employees exceeding 500, and to other entities exceeding the status of medium-sized enterprise and either total asset on the balance exceeding €10 million. The subject matter of reports, following the model of the guidelines, should include information about the organization, its activities and change of the scale of those activities in the longitudinal aspect, a profile of data on substations such as energy, raw materials, pollution and emissions, conformity with the law and the rate of resolutions to the environment-related law, its environment-oriented aims and results intended to secure them, a description of policy employed in managing the ecological risks, harm done to the abiotic and biotic environment and mitigation against their deleterious effects, near environment management, notification of breaches of environmental law and financial penalties. The prevailing criticism of the guidelines is the overwhelming quantity of the required information perceived as difficult to gather and manage. These challenges result from insufficient data management systems, the sensitivity of some of the required information and a lack of standardization of methodologies. These include expansions and substantiation of the criticism about the difficulty of complying with the guidelines and a lack of benefits from the outcomes of the standardized reports. There are also misgivings that the excessive detail of the reports is going to obscure important issues dominating the customary public and media discourses on ecological controversies. In terms of institutional implementation of the standardization so far, there are still no regulations in national legal orders in the majority of European states.

25.3. Predictions for the Future of GRI Reporting

Review: Discusses anticipated developments in GRI reporting and sustainability practices.

Key Topics: Increased standardization, global adoption, and the role of GRI in addressing emerging global challenges.

Outcome: Provides insights into the future direction of GRI and its potential to shape global sustainability reporting standards.

Three main predictions can be made about the future use of the GRI guidelines: (1) Organizations will make more use of Specific Standard Disclosures; (2)

Reporting focus will continue to diversify between time periods; and (3) Organisations will report more on material aspects. (Wuerzinger, 2015) analyzes the predictions by using a dataset including the 297 first-time reports released by existing reporters within the two years following the launch of the GRI G4 Guidelines. Coefficients are run of time since the report's publication on the likelihood to disclose a Specific Standard Disclosure related to a given disclosure category time period or a specific disclosure time period. The results show that between Time Period Negligible and Time Period Relevant reports the additional likelihood to disclose a Specific Standard Disclosure related to a new disclosure category decrease over time.

As concerns the use of Specific Standard Disclosures, it is found that the additional likelihood to make a Standard Disclosure related to a new aspect decreases with the increase in market attention to that aspect. The first of two second-term predictions is found to hold, with an increasing number of Standard Disclosures being released. This implies that the focus on the disclosure of General Standard Disclosures is gradually giving way to reporting on the Specific Standard Disclosures. However, no statistical evidence is found for the second prediction that the disclosure focus will rotate between different time periods over time.

25.4. Challenging Questions and Answers

Q1: How is GRI leveraging technology to enhance sustainability reporting?

A1: GRI is integrating digital tools, AI, and blockchain to improve data accuracy, enable real-time reporting, and enhance transparency in sustainability disclosures.

Q2: What are the key predictions for the future of GRI reporting?

A2: Predictions include increased use of Specific Standard Disclosures, greater focus on material aspects, and broader global adoption of GRI Standards.

Q3: How will GRI address emerging global challenges in sustainability reporting?

A3: GRI will continue to innovate, standardize reporting frameworks, and adapt to address challenges like climate change, resource scarcity, and social inequality.

Q4: What role will standardization play in the future of GRI reporting?

A4: Standardization will enhance comparability, consistency, and credibility of sustainability reports, making it easier for stakeholders to assess and compare organizational performance.

Q5: How can organizations prepare for future trends in GRI reporting?

A5: Organizations should invest in advanced data management systems, adopt emerging technologies, and align their reporting practices with evolving GRI Standards.

Q6: What impact will GRI have on global sustainability practices?

A6: GRI will drive global adoption of standardized reporting practices, fostering transparency, accountability, and sustainable development worldwide.

25.5. Summary

Chapter 25 examines the future of sustainability reporting, focusing on GRI's role in driving innovation and standardization. It highlights emerging practices like digital tools, AI, and blockchain, which enhance reporting accuracy and transparency. The chapter predicts increased global adoption of GRI Standards, with a shift towards Specific Standard Disclosures and a focus on material aspects. GRI's adaptability and leadership are emphasized as key to addressing emerging global challenges, such as climate change and social inequality. By fostering standardization and innovation, GRI is poised to shape the next generation of sustainability reporting, ensuring greater transparency, accountability, and alignment with global sustainability goals.

References

Achmad, T., Faisal, F., & Bahrul Ulum, M., 2017. Nature Of Information Provided And Quality of Sustainability Reports Assurance Statements Of Fortune Global 500 Companies. [PDF]

Addison, P.F.E. & W. Bull, J., 2018. Using conservation science to advance corporate biodiversity accountability. [PDF]

Arvidsson, S., 2018. How are Stakeholder Relations Communicated in Sustainability Reports? A Longitudinal and Cross-country Focus. [PDF]

Australia and New Zealand, A. C. C. A. & Balance Foundation, N., 2013. The state of sustainability report assurance in the ASX 100. [PDF]

Australia and New Zealand, A. C. C. A., Balance Foundation, N., & Investment Research Institute, S., 2013. Disclosures on water. [PDF]

Avrampou, A., Skouludis, A., Iliopoulos, G., & Khan, N., 2019. Advancing the sustainable development goals: evidence from leading European banks. [PDF]

Barkemeyer, R., Preuss, L., & Lee, L., 2015. On the effectiveness of private transnational governance regimes: Evaluating corporate sustainability reporting according to the Global Reporting Initiative. [PDF]

Beghetto, V., Cayzer, S., & Griffiths, P., 2017. Design of indicators for measuring product performance in the circular economy. [PDF]

Berg, A., Antikainen, R., Hartikainen, E., Kauppi, S., Kautto, P., Lazarevic, D., Piesik, S., & Saikku, L., 2018. Circular Economy for Sustainable Development. [PDF]

Calitz, A.P. & F Zietsman, J., 2018. An Adapted Framework for Environmental Sustainability Reporting using Mobile Technologies. [PDF]

Faisal, F., 2017. Nature Of Information Provided and Quality Of Sustainability Reports Assurance Statements Of Fortune Global 500 Companies. [PDF]

Ghoogassian, C., 2015. Evading the Transparency Tragedy: The Legal Enforcement of Corporate Sustainability Reporting. [PDF]

Gibassier, D., Rodrigue, M., & Arjaliès, D. L., 2016. From Share Value to Shared Value: Exploring the Role of Accountants in Developing Integrated Reporting in Practice. [PDF]

Goldman, J., 2015. Integrated reporting : inspiring companies to integrate sustainability into their business strategy and practice? [PDF]

Hysa, X., Uc, M., Gruda, S., & Maliqi, A., 2015. Sustainability Reporting Initiative in Balkans' Higher Education Institutions: The Case of ECOSSS. [PDF]

Mansoor, H., 2016. Biodiversity reporting in South Africa: corporate disclosures in the mining and food sectors. [PDF]

Moraga, G.L., Huysveld, S., Mathieux, F., Andrea Blengini, G., Alaerts, L., Van Acker, K., De Meester, S., & Dewulf, J., 2019. Circular economy indicators : what do they measure?. [PDF]

Parsa, S., Roper, I., Muller-Camen, M., & Szigetvari, E., 2018. Have labour practices and human rights disclosures enhanced corporate accountability? The case of the GRI framework. [PDF]

Pavaloaia, L., Dicu, R., Chelariu, G., & Mardiros, D., 2017. Study on GRI Reporting of Non-profit Organizations in Europe. [PDF]

Romana, L.L.S,, 2018. Measuring the impact of business: Sustainability Reporting by Corporations in Emerging Asia. [PDF]

Smit, A.M. & Magdalena van Zyl, J., 2016. Investigate the level of disclosure of emissions of the top ten manufacturing companies in South Africa. [PDF]

van der Walt, C., 2018. Sustainability reporting practices in small-to-medium sized enterprises. [PDF]

Wu, J., Arthur, C. L., & Yago, M., 2017. Performance Indictors disclosure in Sustainability Reports – Lessons from Ghanaian large Mining Companies. [PDF]

Wuerzinger, J., 2015. "What matters?": The case of sustainability reporting in Spain and Portugal: market response and materiality mapping of the recently launched global reporting initiative's (GRI) G4 guidelines. [PDF]

FSC
www.fsc.org
MIX
Papier | Fördert
gute Waldnutzung
FSC® C083411

Kolibri 360 GmbH
Ferdinand-Jühlke-Straße 7
99095 Erfurt, Deutschland
produktsicherheit@kolibri360.de

Druck:
CPI Druckdienstleistungen GmbH
im Auftrag der
Kolibri 360 GmbH
Ferdinand-Jühlke-Str. 7
99095 Erfurt